D0887113

Evolution In Perspective

Evolution In Perspective

COMMENTARIES IN HONOR OF

Pierre Lecomte du Noüy

GEORGE N. SHUSTER
RALPH E. THORSON
Editors

UNIVERSITY OF NOTRE DAME PRESS
NOTRE DAME & LONDON

Library of Congress Catalog Card Number: 78-105725
Printed in the United States of America by
NAPCO Graphic Arts, Inc., Milwaukee, Wisconsin

CONTENTS

BIOGRAPHICAL NOTES

ALFRED S. BALACHOWSKY was educated at the University of Paris and the University of Rennes, with a diploma in engineering and a doctorate in the natural sciences. His research activities have been concerned primarily with biological warfare against destructive insects. He is a member of the Institut.

PAUL J. BRAISTED is president of the Hazen Foundation and secretary of the Du Noüy Foundation. He was educated at Brown University and Columbia University (Ph.D., 1935) and has had a lifelong interest in religion, especially in terms of international conciliation.

RÉMY CHAUVIN was educated at Laval University and the University of Paris, and holds the degree of Doctor of Natural Sciences. Some of his numerous and distinguished publications treat of insect life. During recent years he has been a member of the faculty of the University of Strasbourg.

OLIVIER COSTA DE BEAUREGARD received his doctorate in science from the University of Paris and is presently with the Institut Henri Poincaré in Paris. His research interests encompass relativity, quantum theory and their mutual relationships.

KARL K. DARROW was educated at the University of Chicago in physics (Ph.D., 1917). He worked in industry with several firms but chiefly with the Bell Telephone Laboratories. He has been a visiting professor at Columbia, Chicago, Stanford, and Smith. He received the Compton Award in 1960. He is now Secretary Emeritus of the American Physical Society.

LOUIS DUPRÉ was educated in Belgium and graduated from the University of Louvain. He is professor at Georgetown University, specializing in the philosophy of religion. Among his works published in English are *Kierkegaard as Theologian, The Philosophical Principles of Marxism, Faith and Reflection.*

DOMINIQUE DUBARLE, O.P., received his doctorate from the University of Saulchoir (1933) and is at present dean of the Faculty of Philosophy at the Institut Catholique de Paris. His published works include *Humanisme scientifique et raison chrétienne* (1953) and *Approche d'une théologie de la Science* (1967).

SIDNEY W. FOX was educated at the University of California and the California Institute of Technology (Ph.D., 1940). He has been a professor and director of the Institute of Molecular Evolution, University of Miami, since 1964. The author of pertinent books in his fields of interest, he is also a well-known contributor to scholarly journals.

CHARLES GOILLOT earned his doctorate in physical science at the University of Grenoble and is now director of the Metereological Laboratory in the Institut National de la Recherche Agronomique. He has contributed numerous papers to scientific journals.

MORRIS GOLDMAN was educated at Brooklyn College, Columbia University, and at Johns Hopkins University (Sc.D., 1953). He served with the United States Public Health Service until 1966 when he joined Bionetics Research in Kensington, Maryland. His research concerns itself with the biology and public health aspects of protozoa of medical importance.

HENRI GOUHIER, Dr. es Lettres of the École Normale Superieure, is Honorary Professor of Philosophy at the Sorbonne and a member of Academie de Sciences Morales et Politiques. He is the author of many books in the history of philosophy and was given the Lecomte du Noüy Prize in 1962.

PIERRE-PAUL GRASSÉ has been professor at the University of Clermont-Ferrand, the Faculté de Sciences in Paris and co-director of the Institute of Psychology of the University of Paris. He is a member of numerous scientific societies in France and other countries. His *magnum opus* is the *Traité de Zoologie.*

CHARLES HARTSHORNE was educated at Haverford College, Harvard University (Ph.D., 1923) and the University of Freiburg. Having held many distinguished academic posts he has been, since 1962, professor at the University of Texas. He is the editor of *The Collected Papers of Charles S. Peirce* and the author of many books, including *The Logic of Perfection,* which received the Lecomte du Noüy Award.

JEAN LADRIÈRE is professor ordinaire at the Université Catholique de Louvain. He has been visiting professor at Duquesne University, Pittsburgh, and the Université Lovanium in Kinshasa, Congo. Besides several translations he has written *Les limitations internes des formalismes* (Paris-Louvain, 1957) and articles on various philosophical subjects, especially in the fields of mathematics and social philosophy.

THOMAS LANGAN is a professor in the University of Toronto, and prior to that was professor of philosophy at the University of Indiana. He was educated at St. Louis University and the Institut Catholique, Paris (Ph.D., 1956). He has written extensively on figures in modern philosophy, notably Heidegger and Merleau-Ponty.

MARY LECOMTE DU NOÜY, translator of many of her husband's books and the author of his biography, was his closest collaborator. She is the founder of the Lecomte du Noüy Foundation.

PIERRE LEROY, S.J., Dr. es Sciences, was during twenty years director of the Center of Geo-Biology in Peking, his colleague being P. Teilhard de Chardin. He is a member of the Centre National de la Recherches Scientifiques.

HENRY MAVIT has higher degrees in literature and law. He is professor in the Institut d'Administration Scholaire and administrator in the Ministry of National Education. One of his books is *Intelligence créatrice.*

FRANÇOIS MEYER received his doctorate from the Sorbonne, and is dean of letters at the University of Aix. His publications include *Teilhard et les grandes dérives du monde vivant.*

GERARD MILHAUD was educated at Zurich and received his medical degree and Ph.D. at the University of Paris (1951, 1954). He is a professor of the Faculty of Medicine, Paris. He works in the area of pharmacology and toxicology. He is a Fellow of the Royal Society of Medicine.

EDOUARD MOROT-SIR, cultural counselor to the French embassy in the United States (1957-1969) and now professor, in the University of Arizona, is an agrégé in philosophy and a doctor of letters. He has held a number of distinguished academic appointments and is well known in particular for his writings about Henri Bergson and Gaston Berger.

GEORGE N. SHUSTER was educated at Notre Dame University, the Université de Poitiers, and Columbia University (Ph.D., 1940). He is president emeritus of Hunter College of the City of New York, and has been since 1961 assistant to the president, Notre Dame University.

RALPH E. THORSON was educated at the University of Notre Dame and Johns Hopkins University (Sc.D., 1952). Having held important posts in academic institutions and in industry, he became professor of biology at Notre Dame in 1959, and, granted a leave of absence (1964-1966), he served on the faculty of the University of Beirut. He is a contributor to the literature of parasitology and is a Fellow of the American Academy of Microbiology.

JACQUES TREFOUËL was educated at Paris and has served as the director of the Institut Pasteur (1960-1964). His scientific interests are in the area of chemotherapy of syphillis, trypanosomiasis, malaria and other diseases.

PAUL WEISS was educated at the City College, New York, and at Harvard University (Ph.D., 1929). Associated with Bryn Mawr College for many years, he joined the faculty of Yale University and became Sterling Professor of Philosophy in 1962. His many books include *Modes of Being* and *The God We Seek.*

RALPH W. G. WYCKOFF was educated at Hobart College and Cornell University (Ph.D., 1919). After a distinguished career as a pioneer in crystal physics and biochemistry, he joined the faculty of the University of Arizona as professor of physics and bacteriology. His many contributions to scholarship include the numerous editions of *The Structure of Crystals.*

PREFACE

This book brings together papers and comments read during a conference and a *colloque* which discussed "Evolution in the Light of Contemporary Research and Reflection." The conference held at Notre Dame University on October 23, 24, and 25, 1967, was international in character. Distinguished scholars from France and Belgium participated, often at a considerable sacrifice of time and energy. There was also participation by scientists, philosophers, and theologians from other countries. The *colloque*, which convened in Paris at a somewhat earlier date, was an exclusively French event. Yet it was also interdisciplinary in character.

Both meetings presented original thinking about the nature of man and the dimensions of the efforts recently made to describe that nature more adequately. They also commemorated the life and work of Pierre Lecomte du Noüy, whose contributions to the advancement of science were as notable as was his thinking about the direction which science might take in the humanistic sense. *Human Destiny* was the first among relatively contemporary efforts to understand man not as a biological phenomenon only but also as a truly human being. Published in 1947, during a time when a number of similar essays were written, it was read by millions the world round and continues to be greatly esteemed as the embodiment of belief that our common purpose as men and women is to transcend purely animalistic behavior.

At this point we may note that there is a very close resemblance between the thought of Teilhard de Chardin and that of Pierre Lecomte du Noüy. They both agreed that the story of life on our earth can be told only in evolutionary terms, that the human brain is the summit of the evolutionary process, and that the future history of mankind will depend upon how that brain functions.

But important differences between the two must be taken into account. Chardin was almost Hegelian in his emphasis on both the collectivity of human experience and the collective-corporate future of mankind. For Lecomte du Noüy, on the other hand, responsibility for moral progress and intellectual development rests with the individual. Moreover, the two were religious men in that they believed the human being has a spiritual destiny, as well as in the more specific sense that they were drawn to the person and mission of Jesus. While the Jesuit priest had spent his life with the remembrance of the Savior in his heart, the scientist moved a long way from an attitude of skepticism about all the claims of religion. Lecomte du Noüy's conversion was based on his discovery that the moral barbarism of his time could not be overcome unless there was at the core of the human person a perception of the Good, which was a response to transcendent Goodness. *Human Destiny* cannot be accounted for without taking into account the revelation of evil by the Nazi uprising and the conflicts it engendered. The First World War, during which Lecomte du Noüy served in an army hospital, was viewed by all who suffered under its impact as a terrible tragedy. Still there was a measure of nobility in it. Men believed that "it is good to die for one's country," and that being called upon to do so was honorable. But the second war raised the question: if man is only a beast, a chance by-product of an evolutionary process, why should he not act like one?

Accordingly the present book is scientific, philosophical, and religious in character. The central theme is of course evolution, seen as the mode in which all three kinds of thought must henceforth be carried on. The scientific papers concerned with evolution—those by Sidney W. Fox, Ralph E. Thorson, Pierre P. Grassé, and Morris Goldman—are quite different in character but nevertheless similar in the sense that they challenge widely held views. Fox sets for himself the task of replacing the nineteenth century assumption that "spontaneous generation" produced life with one that man can start an evolution by reproducing through experimentation the steps by which a living organism presumably came into being. He also provides a brief summary of the laboratory processes used. Grassé's paper is based on a transcription of a rather extempore discourse in which he questions statements made by those he terms "super-Darwinians." In other words, it is a critique of the more extreme supporters of the natural selection thesis. Thorson proceeds to show that contrary to many popular beliefs evolution reveals a remarkable parsimony in nature—in short, that the process of development is simpler than has often been assumed. Goldman's paper contends that the evidence supporting the view that man was produced by a series of orderly progressions from lower mammals is open to serious question. It supports the thesis of theistic evolution by implication, but holds that neither this nor the view that amoral nature can account for moral man can be proved scientifically. It is no doubt the most controversial of the scientific papers, but it was discovered that refuting it was no simple undertaking.

Finalism, which played so important a role in the thinking which underlay *Human Destiny* and the more recently published volume of essays entitled *Between Knowing and Believing*, is the theme of most of the philosophical papers. In his treatment of it, which is also an edited transcription, Rémy Chauvin says:

> Let us content ourselves with saying that in living beings and particularly in man, it seems that we can ascertain the existence of an activity directed to a goal that remains constant, through the multiplicity of situations and perturbations encountered. It seems as if there were an aspect of finalized activity which, starting from man, is extended to all living beings and therefore to the whole of evolution.

This is of course a tentative definition but it will serve our purpose.

Two of the most notable papers in this book deal with finalism in different ways. That by Jean Ladrière is a highly original attempt to carry the notions of freedom and finality beyond Kant. Near the close of his essay, he says:

> But if there is, at the very source of genesis, freedom as a positing act, then the meaning of our question about the finality of nature is modified. The question of finality turns into that of meaning. It is no longer about the action of the totality on its parts nor even about the possibility of nature's going beyond itself. Rather it is about the arrival in nature of a meaning which nature knows nothing about and the secret of which is possessed only by the positing act. Thus the true finality of nature perhaps lies well outside of it.

The second paper, by François Meyer, presents a careful analysis of the concept of finalism which Lecomte du Noüy expounds in *Human Destiny* and elsewhere. He gave this the name of telefinalism, in order to distinguish between adaptation and evolution.

The other philosophical papers are historical in character. That by Edouard Morot-Sir is concerned primarily with the philosophy of science. He says:

> In other words, a scientific result, no matter how objective it may be, has if it is considered in its totality a historical and a national dimension.

Morot-Sir then proceeds to discuss the role of Cartesianism in French thinking about science, the differences between logic and mathematics, and the possibility of "a universal theory of science." Henri Gouhier deals with the differences between the history of philosophy and the history of science. He speaks of "philosophical exigence," and in this context discusses what the historian of that exigence does and the sources from which he draws. He says:

> It is Pascal, discovering also the convergence of a new science . . . who makes us pass from a closed world to an open world, who discovered the convergence of a science which projected an indefinite world with a theology which adores a hidden God and forces us to question even the words of a psalm.

The religiously oriented papers evoked the most comment and discussion, doubtless because no commemoration of Lecomte du Noüy could avoid stressing the spiritual outlook which characterized the later period of his life and work. The French Dominican Dominique Dubarle is a theologian, a philosopher, and a scientist. But far from making his thought a series of generalizations, this trinity of interests gives it extraordinary riches and depth, as readers will discover for themselves. While remaining grateful to Lecomte du Noüy, he subjects some of his philosophical assumptions to scrutiny. But his concern is really with the "lyricism" of Teilhard de Chardin, to which he opposes the possibility of a negative outcome of the evolutionary process insofar as the spirit of man is its supreme outcome. He also stresses the seeming indifference of the universe as we are capable of knowing it to the emergence of life itself.

Paul Weiss has a remarkably subtle mind. His paper expresses dissatisfaction with what is basically Cartesian in the philosophical outlook of Lecomte du Noüy, dissents from the view that science and religion are the only fields of exploration and inquiry open to the philosopher (others being, for instance, art and politics), and concludes:

> In summary, I think we must say, that the world contains a plurality of substantial beings which are affected by transcendental realities to turn them, among other things, into entities in a cosmos, or into entities which have a sacral stature, the one being studied in science and the other accepted in religion.

Charles Hartshorne's comment could not be presented at the Notre Dame Conference, regrettably enough. But it is perhaps desirable that this was the case, because the comment takes into account not only Weiss's paper but also his relevant books, as well as the writings of Lecomte du Noüy and Teilhard de Chardin. In a key passage concerning the relationships between science and religion he says:

> Am I denying that "the heavens declare the glory of God"? No, for to the believer they do declare it—but with the understanding that any conceivable heavens or absence of heavens would do so for a sufficiently wise observer. The eminent cause must be in any effect; if it is not seen there, so much the worse for the penetration of the seer. The special character of this actual world, which is what science seeks to reveal to us, shows us not that God exists (for he would do so in any possible case) but only what universe he contingently has. Apart from science we could know that God, the eminent one, exists, but what sort of a world he has we could not know.

Louis Dupré's paper and Thomas Langan's comment on it restate the problem of how a divine causality can be made compatible with the final product of evolution, *free man.* Dupré says, "Lecomte du Noüy epitomizes this in the paradoxical expression, 'God abdicated a portion of his omnipotence when he gave man liberty of choice.' " Dupré finds in the philosophy of Henry Duméry a significant approach to a new and fruitful discussion of the freedom which defines man. "If," he writes, following Duméry, "the prescientific concept of unlimited freedom is too simplistic, the positivist concept of an all-comprehensive preexisting determinism is even more so." The paper closes on the note which was also final in the thought of Lecomte du Noüy, namely the acceptance of faith in Christ. Thomas Langan's comment outlines some of the difficulties he sees implicit in Duméry's position. These rise out of conflicts between an emanationist trend to be offset by one that is phenomenological in character.

II.

The second part of this book is dedicated to the life and work of Lecomte du Noüy, though to be sure what precedes it makes many a pertinent reference to him. What follows offers by far the most extensive available commentary on his personality and his tireless dedication to science, though it needs supplementing with the biography written by his wife. Two scientific papers, by Costa de Beauregard and Gerard Milhaud, appear here because they deal with problems in which Lecomte du Noüy was specially interested. Everyone with cherished memories has also been allowed to speak his piece.

Ralph W. G. Wyckoff's paper is not only a tribute by a distinguished biochemist to the memory of another, but it is also an impressive restatement, in terms contemporary with ourselves, of the basic problems with which Lecomte du Noüy was concerned. To be sure, the work of a pioneer in a science is superseded by that of those who succeed him. All one can do is to recall the originality of his insights and the rigor of his demonstration of their value. But there is a more basic question. How does one look for meaning in life, not as an individual person only but as a member of that race on the earth which alone can lay claim to intelligence?

Wyckoff, continuing the thought of Lecomte du Noüy, says that the fruits of scientific discovery and the joy in helping to gather them may be so great that nature will satisfy the scientist so that he will not feel a need for the transcendent. Thus materialism has been and still is the creed of many. But it will seem insufficient to some, as it did to Lecomte du Noüy. Wyckoff writes:

> We can observe in ourselves and the higher animals how the gradual development of consciousness transfers these reactions and decisions progressively from the instinctive to the willfully determined. We also see their dual character: some are based on the material realities, the truth, or situation, others on desirability.
>
> In the conscious life of man the rational intellect represents our way of coping with questions of truth, while our criteria of desirability have evolved out of purely biological urges to include a more or less disciplined sense of esthetic and spiritual values. The history of human culture is the story of the gradual development of these two components of that inner life which the emergence of consciousness has made possible.

He concludes by conjuring up the quite awe-inspiring responsibility which rests on man to see that evolution is oriented towards the Good.

Gerard Milhaud's paper is concerned with the aging process and the possibility that it could to some extent at least be slowed down. This process was one which interested Lecomte du Noüy from the beginning of his career as a scientist. Does not the body have a built-in time clock which ticks away until the hour of death? It cannot be turned back. So far medical science has succeeded in removing or retarding external deleterious forces which would stop the clock prematurely—tuberculosis, polio, small pox, for example. But despite some interesting experimentation nothing has been discovered which affords the hope that the process itself can be regulated. Milhaud quotes Chamfort to the effect that our choice appears to be *Être passioné et vivre ou être raisonnable et durer.*

It is impossible to summarize Costa de Beauregard's closely reasoned paper, which deals with two major aspects of time. First, he says, "it is a measurable magnitude as physicists say in their jargon." To be sure, the definition of the magnitude has been reformulated as science proceeded from Galileo to Einstein and Minkowski. Second, a "major aspect of time in physics is the irreversibility occurring whenever the energy balance implies heat—that is, in such commonplace phenomena as friction or such unsophisticated devices as brakes, as well as in the much more sophisticated heat engines obeying the famous Carnot-Clausius law." Starting with these definitions the author proceeds with his discussion of the "irreversibility problems," which in his view bring physics close to metaphysics.

Charles Goillot's paper is much more directly concerned with Lecomte du Noüy, dealing as it does with the instruments he designed to assist in experimentation. He himself wrote:

> A survey of the advancement of knowledge brings the realization that the progress of science has been achieved through the constant improvement of experimental techniques. This of course does not apply to the immense strides made by the speculative genius of men like Newton, Maxwell and Einstein, who belong in a class apart.

Of course, the paper is no mere enumeration and description of the instruments. It is also a commentary on the uses to which they were put and the results which were achieved.

Karl Darrow's paper is a genial summation of Lecomte du Noüy's career as a scientist, and it is based partly on personal memories and on the *Exposé des Travaux de Pierre Lecomte du Noüy*, published in 1937. The author concludes by saying:

> There is an old and very stale joke about the university dean or president who evaluates the members of his faculty by counting or weighing their publications. This joke almost frightens me away from giving the statistic of Pierre's papers. But I do not allow myself to be frightened, for it is a valuable and meritorious fact that in these years Pierre published 112 papers and 6 books.

Our papers commemorative of Lecomte du Noüy must of course be introduced with the extraordinarily moving and informative talk by his wife, given at a small dinner arranged at Notre Dame for the occasion. It is not merely reminiscent but evocative of the progress of science during the decades of their common life together. Perhaps one of the most touching of her remarks is this:

> *Human Destiny* rapidly made the best seller list, has now attained the highest number of editions of any work of its kind, and has been translated into twenty-three languages. Lecomte du Noüy did not live to know its tremendous success throughout the world. After my husband's death, Millikan—who never passed through New York without coming to see me—repeatedly told me he always kept *Human Destiny* beside his bed, right next to the Bible, and referred to it constantly.

No doubt this communion between scientists who considered not only the advancement of science but the development of spiritual insight is the most important achievement of the man whose memory was stressed in both the conference and the *colloque*.

There are included in this book several personal (and, one might add, very French) outlines of reminiscence. Those of us who know and love France have found out what *amitié* means. Lecomte du Noüy must have been a really wonderful practioner of that art, which may well be the greatest, though currently perhaps the least appreciated, of all the arts.

The papers by Jacques Trefouël and A. S. Balachowsky are precious memoirs. They speak of a gifted and radiant friend and of the harrowing time of the agony of all the world. They will reinforce in those who read them the commitment to what is not tawdry but is comprised of abiding values which are so human because they provide man with a basis for aspiration. These memoirs are supported by the other briefer papers–those by Paul Braisted, Pierre Leroy, S.J., and Henri Mavit.

The conference and the *colloque* would not have been possible without the generous support of the Lecomte du Noüy Foundation, which has been established both in France and the United States. The principal purpose of the foundation is to honor with a prize the author of a book dealing with some aspect of the relationship between scientific activity and spirituality. But it is now seeking to broaden its field of interest. One of its major concerns will henceforth be to foster ties between scholars in this country and in France.

George N. Shuster
Ralph E. Thorson

Can Man Start an Evolution?

Sidney W. Fox

In the preface of *Human Destiny* Lecomte du Noüy said in 1947, "We beg the reader to make the effort of 'handling' the ideas which are not familiar to him by criticizing them, by taking them to pieces, and by trying to replace them by others."

Most memorable, I believe, is what Lecomte du Noüy has done by way of identifying areas of what proved subsequently to be significant content—for example, entropy and evolution, protein isomerism, and the origin of life.

One idea that I believe should be replaced, as stated by Du Noüy, is the notion that man will never be able to start an evolution. Consistent with the central postulate of a theory of spontaneous organic synthesis proposed by Oparin in 1924, experiments since 1950 have shown at almost all levels that the components and structure of life can arise spontaneously in the absence of life. The degree to which we can defensibly aver that, with this as a base, man can start an evolution is a judgment that requires much expert and willing scrutiny for at least several hours, preferably on separate occasions. I shall be able to present evidence for less than forty-five minutes. However, even a spotty review of the evidence can make clear that we cannot state without qualification that man will certainly never be able to start an evolution.

Accordingly, I am taking seriously the admonition in the preface to *Human Destiny* of trying to replace one of the main ideas therein. Relevant to this and other questions, I again quote Lecomte du Noüy. "New facts are frequently found in science which compel him (man) to revise completely his former concepts."

The problem of the origin of life, or in truly perceptive nineteenth century terms, the problem of spontaneous generation, has often been regarded as one of overwhelming complexity. Upon analysis, with the aid of hindsight, this problem loses some of its imponderability. The aspect of evolution which first received major attention was that of the progression, in principle, from primitive cell to contemporary cell and to contemporary multicellular organisms. This stage is the one that has been illuminated mechanistically by Darwin's theory of selection. We can now regard this stage as far more intricate and involved than the emergence of primitive life from the primordial reactant gases. By such an analysis, the primordial cell is emphasized, the highly ramified later stages are removed from purview, and the limits of the meaningful problem are identified.

The preorganismic stage can also be analyzed. For intellectual convenience, it may be divided into two or three parts. The first of these parts is that of the spontaneous organic synthesis involved in the production of the small organic molecules which are necessary for contemporary and, presumably for, primitive organisms. The second step is the spontaneous synthesis of the polymers and of cells. This latter constitutes in turn, however, two stages. These two steps were collectively most forbidding in quality, and are particularly significant to life and therefore, to its origin. Our most modern knowledge requires that we recognize that a primitive cell cannot be a synthesized entity in the true meaning of "synthesized." The precursor macromolecule can be conceived of as synthetic. When the appropriate macromolecule has been formed, the final and crucial stage, leading to a primitive organism, would then be one of self assembly. The term "self assembly" and the concept have recently been receiving increasing recognition, e.g., in the biochemistry of contemporary systems.[1]

One way in which students of the total problem have dealt with the seemingly great complexity has been to postulate a long chemical evolution extending over, say, 25 million years.[2] I will explain here why our experiments lead to the interpretation that the essential steps from primordial gases → amino acids → primitive protein → a primitive organized structure having simultaneously many lifelike properties including the ability to participate in its reproduction could have occurred many times in a very short period, perhaps twenty-five hours. My immediate problem is to present the salient experimental material in twenty-five minutes. This problem exists because of the careful devotion to it by many associates during fourteen years of continuous study in our laboratory. Accordingly, I shall rely heavily on summaries and upon examples from our laboratory and others.

Our approach to this problem was based on clues from contemporary cells. The results of experiments have been evaluated in part by how well they lead to an increasing appearance of the properties that are associated with contemporary cells. However, the experiments have been based on very simply derived initial systems and simple processes. These employ conditions that have proved to be plausible not only for geologically ancient times. The conditions identified are widespread now and through recorded history.[3]

Models of the prebiotic synthesis of small organic compounds such as monosaccharides, amino acids, purine and pyrimidine bases, ATP, porphyrins, etc. have been described from many laboratories including those of Ponnamperuma,[4] Oró,[5] Oxgel,[6] and our own.[7] Since the essence of life is generally recognized as being that of the biopolymers, protein and nucleic acid, this paper will focus on questions involving the primordial formation of protein and nucleic acid and on the attributes of the polymers formed in the laboratory. It will deal also with complexes of the two.

Turning first to the question of proteins, we find that a number of studies of the synthesis of peptide bonds, mostly in aqueous solution, have been carried out in a number of laboratories. Akabori employed the progressive substitution of polyglycine as a model of the first protein,[8] and Matthews has reported a similar process.[9] The model of our laboratory which relies on heat and hypohydrous conditions is the only one that has yielded polymers of molecular weight in the thousands, a content of all eighteen amino acids common to protein, several protoenzymic activities, and it is the only model which has been demonstrated to yield organized structures with a lengthy roster of the properties of the contemporary cell.[10] This synthesis has the simplicity appropriate to geologically spontaneous occurrences, and it yields both polymer and organized units in abundance.

One other synthesis, which has most of the attributes enumerated, uses as intermediates the reactive Leuchs anhydrides of the amino acids. This synthesis was also first performed in our laboratory by Dr. T. Hayakawa.[11] Of these two syntheses, only the thermal process has the simplicity appropriate to geologically spontaneous occurrences.

The thermal syntheses first attempted in 1953 were indicated as thermodynamically possible following studies of Borsook, Huffman, Ellis, and Fox.[12] The results of calculations from the tabulated physical constants have shown that one could expect in an open aqueous solution only small yields of small peptides unless the reaction were somehow coupled to an endergonic one.

The fact that organisms are nearly always aqueous entities has led some to assume that hot, dry conditions would not have been appropriate to early life and they have somehow projected such thinking to precursor molecules. Our chemical experiénce, however, tells us that macromolecules can easily survive conditions lethal for ordinary cells, and our biological experience reminds us that dry bacterial spores are relatively highly resistant to heat and dryness.

The reaction involving formation of peptide bond with its attendant Gibbs free energy change is:

$$H_2NCHRCOOH + H_2NCHR'COOH = H_2NCHRCONHCHR'COOH + H_2O$$

$$\Delta G^\circ = 2000 \text{ to } 4000 \text{ cal.}$$

As the number of peptide bonds per molecule increases, the equilibrium constant becomes geometrically more unfavorable. Dixon and Webb[13] have calculated that the volume of 1 m amino acid solution in equilibrium with one molecule of protein of molecular weight 12,000 would be 10^{50} times that of the earth. Stated otherwise, uncoupled synthesis from amino acids in water should be expected to give small yields of small peptides only.

In order to shift this equilibrium to favor synthesis, one can postulate removal of either product. Theoretically, one contribution of a membrane in contemporary protein-synthesizing systems may be the overcoming of an energy barrier by separation of synthesized macromolecules from the aqueous solution. This process could not apply, however, until a membrane composed of macromolecules had first formed. Our attention therefore shifts to removal of the other product, water. This route to peptide bond synthesis can be visualized as a geochemical possibility. It has also been demonstrated experimentally.[7]

One mode of removal of water, as thus suggested by the thermodynamic analysis, would be that of heating the amino acids above the boiling point of water. When this possibility was initially contemplated, the probability of gross decomposition had to be considered. Such a consequence of heating α-amino acids above the boiling point of water has been recorded in the literature a number of times and was also common knowledge. We were led to attempt the thermal condensation by employing an inference from comparative studies of organismic protein, the fact that the amino acids which most dominate the composition of proteins are glutamic acid and aspartic acid.[7] These contents were taken hypothetically to be an evolutionary reflection of a circumstance required for the primordial formation of prebiotic protein.

Another consideration that had to be dealt with was the somewhat vague feeling that without nucleic acids present, the necessary systematic sequences of amino acid residues would not result. This problem was conceptually eliminated in principle by studies of enzymic acylpeptideanilide synthesis which demonstrated that interactions of amino acid residues would alone select the sequence formed.[14] (This principle and the inference that prior nucleic acids may have been unnecessary[7] has since been corroborated by Steinman in another system of reacting amino acids.)[15] Since the difficulties were thus conceptually surmountable in 1953, heating was employed (Figure 1). The discussion now deals with experimental observations.

FIGURE 1. On left, tube containing mixture of amino acids heated to above the boiling point of water. On right, granular polymer prepared by heating a mixture of amino acids containing sufficient proportions of aspartic acid and glutamic acid.

A typical thermal condensation used at first a mixture of one part of aspartic acid, one of glutamic acid, and one of an equimolar mixture of the sixteen other amino acids common to protein. This mixture was heated at 170° for six hours.[16] The resulting light amber glassy product, not depicted, is entirely soluble in water by salting-in, and can be purified by salting-out. Such products yield amino acids 100 percent by acid hydrolysis, they contain some proportion of each of the amino acids common to protein (or fewer as desired), and molecular weights of many thousands. They have many other properties of protein and are called proteinoids. The proteinoid described in this example is, because of the proportions reacted, a 1:1:1 type. More recently, Dr. Waehneldt has shown in our laboratory that aspartic acid and glutamic acid may be merely equimolar with the sixteen other amino acids. Proteinoids are produced even so. Yields in the usual syntheses are typically in the range of 10–40 percent, higher yields being obtained by the addition of phosphates.[17,18] Many other laboratories have repeated this synthesis and have confirmed it and its simplicity, which is in turn crucial to the geological validity. The spontaneous occurrence of a carbo-benzoxy synthesis or the formation and condensation of N-carboxy amino acid anhydrides cannot, of course, be defensibly imputed to the geological environment.

The numerous properties which the proteinoids have in common with proteins are described in detail in the literature, and these have been reviewed a number of times.[7,10]

TABLE I

PROPERTIES OF THERMALLY PREPARED PROTEINOIDS

Limited heterogeneity
Qualitative composition identical to that of protein
Quantitative compositions resembling those of proteins
Quantitative recoverability of amino acids upon hydrolysis
Range of molecular weights like those of smaller protein molecules
Positive color tests as for protein
Solubilities resembling those of protein classes
    (albumins, globulins, etc.)
Some optical activity
Tendency to be salted-in
Tendency to be salted-out
Precipitability by protein group reagents
    (phosphotungstic acid, etc.)
Hypochromicity
Infrared absorption maxima as found in protein
Some susceptibility to proteases
Nonrandom distribution of amino acid residues
Many catalytic activities
Inactivatability of catalytic power
Nutritive quality
Melanophore stimulating activity
Morphogenicity

Table I lists the many simultaneous structural, chemical, and biological properties all of which are described in literature cited bibliographically, except for recently demonstrated hormonal activity and a few other aspects. Time will be devoted here to reviewing only three salient properties on which many new data are available: first, limited heterogeneity and the related question of systematic sequences; second, catalytic activity; and third, the property of forming regular and highly structured particles.

The first indication that a thermal condensate of eighteen amino acids yields no more than two electrophoretic individuals was from a study by Dr. Carl Vestling.[16] Subsequently, Dr. Harada in our laboratory showed that two fractionations of a 2:2:3-proteinoid from hot water resulted in virtually no change in amino acid composition.[19] Also, N-terminal[16] or C-terminal analyses[20] of many polymers demonstrated marked disparities from the total compositions, indicating that systematic nonrandom sequentialization occurred. This could be due only to selective interactions of amino acids during thermal condensation. Since that finding, many new data obtained by Dr. Nakashima and a review of all of the data have been published.[21]

The elution pattern from fractionation on DEAE-cellulose is shown in Figure 2.

While no random assortment of polyanhydro-α-amino acids has been prepared such that a comparison would be possible, we would

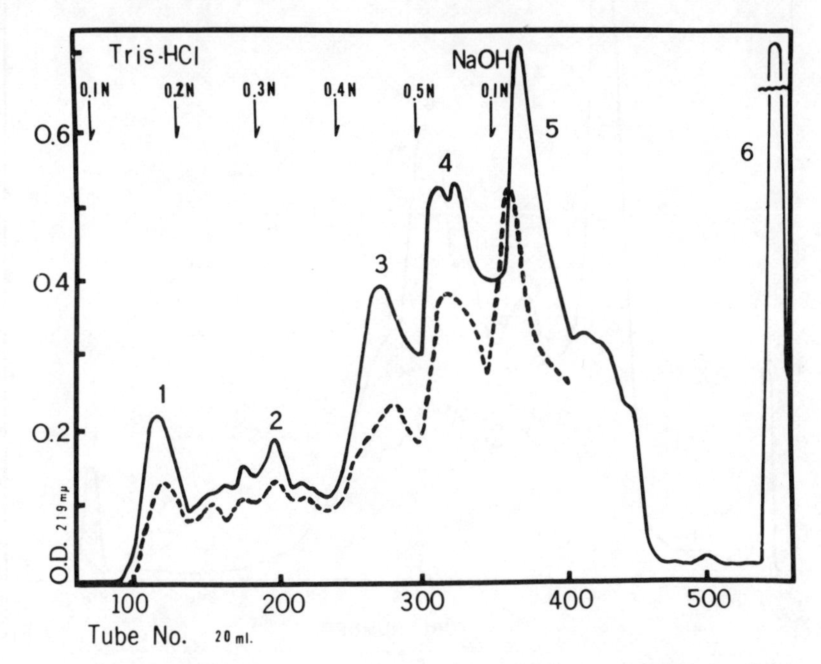

FIGURE 2. Distribution of 1:1:1-proteinoids on elution from a DEAE-cellulose column by tris-HCl buffer.

theoretically expect an elevated nearly horizontal line for an elution pattern of a disordered polymer. What is repeatedly found, instead, is a pattern of six major peaks, of which some are already symmetrical.

For comparison is presented an elution pattern of turtle serum protein[22] also fractionated on DEAE-cellulose. Eight major peaks, with less spread in each peak, are observed. The individual peaks of the eluate sediment in the Spinco Model E (Figure 3) to indicate for the fractions the degree of ultracentrifugal homogeneity which is observed.

Figure 4 shows the complete and partial hydrolyzates of three fractions from DEAE-cellulose. The bottom three are partial hydrolyzates in each group. The patterns are highly similar, as can be seen.

A total picture of nonrandom sequences in the linear distribution, discrete macromolecular fractions, and relatively uniform composi-

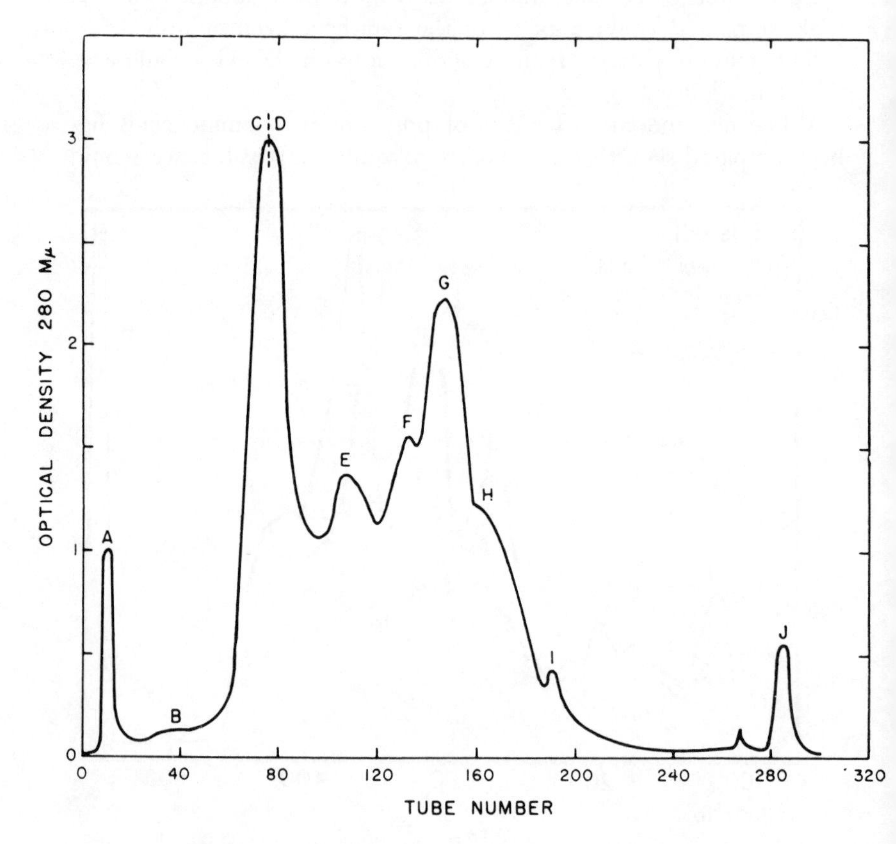

FIGURE 3. Distribution of turtle serum proteins on elution from a DEAE-cellulose column by sodium phosphate buffer.

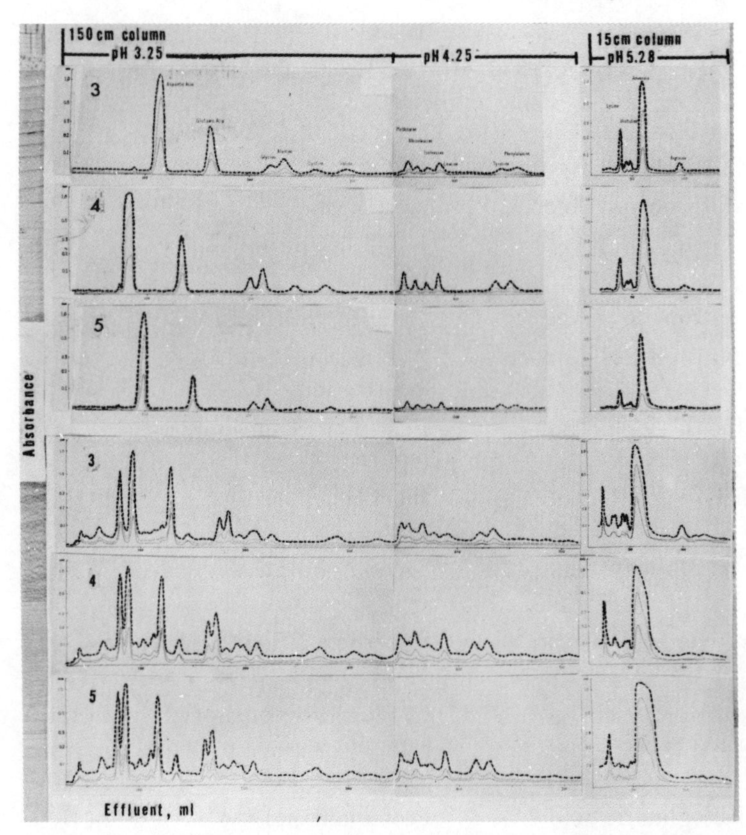

FIGURE 4. Top three profiles of hydrolyzates of three fractions of proteinoid-amide show great similarity in amino acid content. Bottom three profiles show similarity in peptides obtained on partial hydrolysis.

tion and sequences throughout the entire polymer is supported also by high voltage electrophoresis, by separation on columns of Sephadax, by polyacrylamide fractionations, and by gel electrophoresis.

Reports of catalytic activity in proteinoids are summarized in Table II. These findings are from six laboratories and they include catalysis of the hydrolysis of esters and of ATP and decarboxylations of a number of natural substrates.[23] Krampitz has recently recorded an example of transamination. Michaelis-Menten kinetics have been reported in several of the studies. The activities are mostly weak. As Calvin pointed out for the iron-containing enzymes[24] and as I pointed out in 1953,[25] weak primitive enzymic activity would be selected and enriched by Darwinian processes in organisms. Of relevance, also, is the fact that some gross specificities have been identi-

TABLE II
CATALYTIC ACTIVITIES IN PROTEINOIDS

Reaction	*Salient Finding*	*References*
p-Nitrophenyl Acetate	$\text{Activity}_{\text{ptd}} > \text{Activity}_{\text{hsd}}$	(26)
p-Nitrophenyl Acetate	Activities	(27)
p-Nitrophenyl Acetate	Inhibition by organic phosphates	(28)
p-Nitrophenyl Acetate	Detailed treatment	(29)
p-Nitrophenyl Acetate	"Active site," and inactivation	(30)
Glucose → glucuronic acid → CO_2	First natural substrate reported	(31)
ATP → ADP, *et al.*	Biochemical energy source	(32, 33)
p-Nitrophenyl phosphate	Second phosphate hydrolysis	(34)
Pyruvic acid → acetic acid + CO_2	Decarboxylation M-M kinetics	(35, 33)
Oxaloacetic acid → pyruvic acid + CO_2	Catalyzed by ptds of type not active on pyruvic acid	(36)
c-Ketoglutaric acid + urea + glutamic acid	Proteinoid and Cu each needed	(37)

fied and that individual proteinoid preparations each have an array of catalytic activities. A metabolic sequence has been recorded, for example, for oxaloacetic acid → pyruvic acid → acetic acid + CO_2.

We can explain the evolutionary potential of polyanhydro-amino acid catalysts by the fact that in the same macromolecule are found not only a variety of chemically functional groups, but also the products of interaction of the fields of force of two or more of these groups. This picture of chemical polyfunctionality is also, of course, applicable to proteins and emphasizes what A. E. Needham[38] has referred to as "the uniqueness of biological materials." Proteinoid is in one sense perhaps even more unique than any one protein in that it is less specialized, and recalls thereby for serious consideration the concept of "urprotein."[39]

The property of forming structurally organized units on contact with water is crucial to a comprehensive theory of the origin of the first cell. This tendency is intrinsic to many thermal polymers of amino acids, as we reported in 1960. The need for a macromolecular precursor of the first cell has been emphasized in theoretical discussions by Oparin,[40] by Wald,[41] by Lederberg,[42] and by others. The degree to which thermal proteinoid meets this need by providing organized microscopic units having numerous associated properties such as are found in contemporary protein and in contemporary cells could not have been predicted. Only some of the more salient properties will be presented here. Others are documented with references to supporting literature.[43] The properties include, among others, a cellular type of ultrastructure, double layers, abilities to metabolize, to grow in size, to proliferate, to undergo selection, to bind polynucleotides, and to retain some macromolecules selectively. The structures of Figure 5 are usually produced merely by heating the proteinoid with water or aqueous solution. The clear liquid, on cooling, deposits huge numbers of microscopic units, 0.5–80μ in diameter, of quite uniform size in any one preparation. These are usually found as spherules, as in this photomicrograph, but they occur also

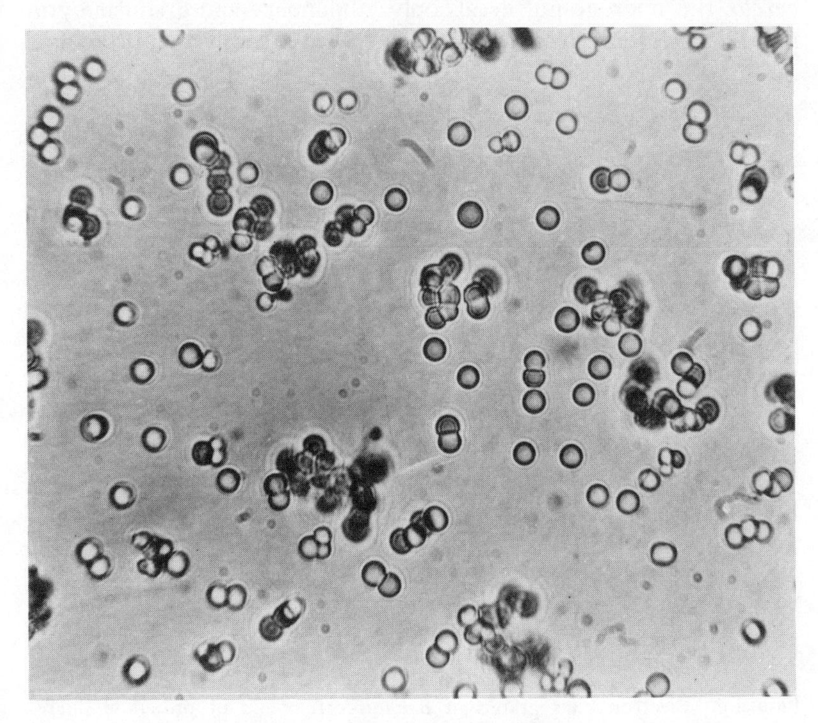

FIGURE 5. Protenoid microspheres.

as filaments, budded microspheres, as twinned units, and in other shapes. They are exceedingly numerous; one gram of heated amino acids can produce many billion spherules. As pointed out by us and by others,[44] they have physical stability comparable to that of contemporary cells; they can, for instance, be sectioned for electron microscopy. In their uniform size and in other respects they are readily distinguished from oily droplets or from the usual coacervate droplets. While they can be produced as entirely separate units, they also tend to associate.[45] By adjustment of the basic amino acid content, they can be produced to stain either Gram-negative or Gram-positive.[46] They also have been shown to have some of the catalytic activities which have been carefully identified in the polymer of which they are composed.[23]

The ultrastructure of the microsphere is shown in the electron micrograph of Figure 6. On the left is a section of *Bacillus cereus* which has been fixed by osmium tetroxide, sectioned, and electron micrographed, by Murray.[47] On the right is seen a section of a proteinoid microsphere fixed with osmium tetroxide and sectioned. While some bacteria reveal a more organized pattern than *Bacillus cereus*, this micrograph reveals only a boundary and granular cytoplasm. The granular appearance is found also in the proteinoid

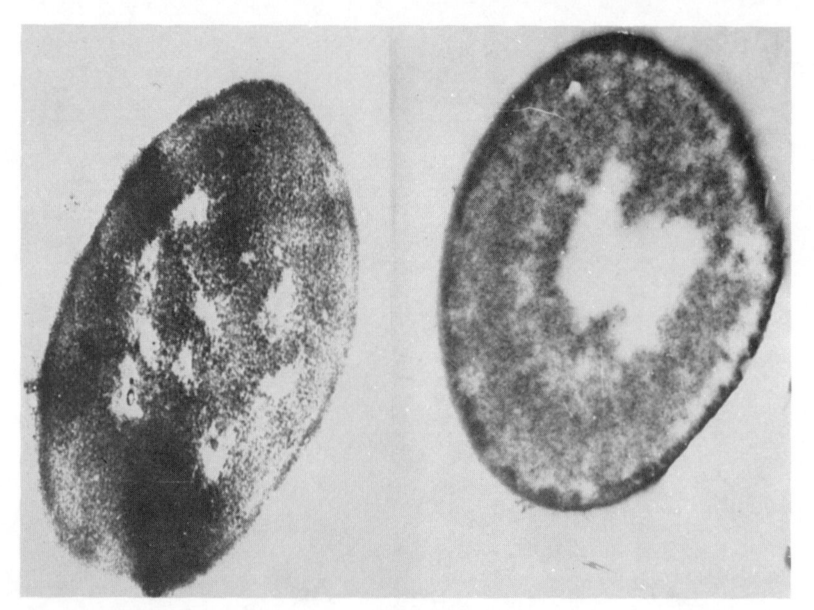

FIGURE 6. Electron micrographs of *Bacillus cereus* and of protenoid microsphere (right). Each has been fixed with osmium tetroxide and sectioned.

microsphere, and the latter has a more definite boundary. In one place the boundary appears to be a double layer. In the experiment preceding the electron micrography of Figure 7, the polymer in the interior was first caused to diffuse out through the boundary by raising the pH in the suspension of proteinoid microspheres by one to two units. Double layers are clearly evident.[7] We can also see part of the results of other phenomena. The diffusion depicted is one manifestation of selective behavior in the boundary or membrane. This diffusion is further illustrated in Figure 8, in which the effect was followed by photographing in ultraviolet light through the quartz optics of Dr. Philip Montgomery's microscope at the University of Texas.

The proteinoid, like protein,[48] absorbs at ultraviolet wavelengths. These pictures and related data indicate that polymer is not condensing on the membrane but is passing selectively through a membrane composed of very similar polymer, as has been shown by analysis.[49]

Models of primitive polynucleotides have also been examined. With Dr. Waehneldt we have reported how nucleoside mono- and

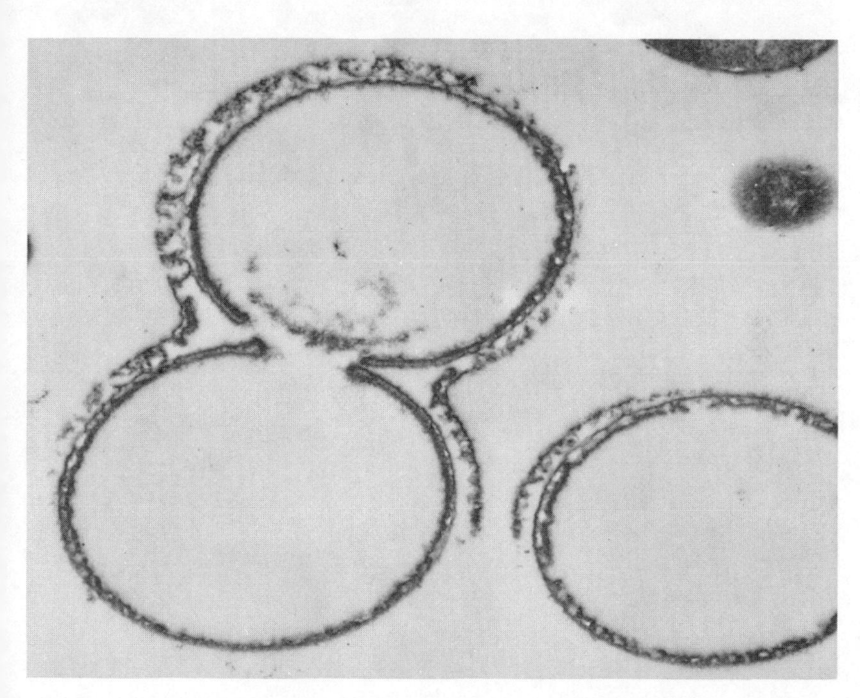

FIGURE 7. Protenoid microsphere subjected to raised pH. Double layers are evident.

tri-phosphates could have been prebiotically synthesized in quantity,[50] following an earlier report on the production of ATP by Ponnamperuma.[51] Thermal polymers of mononucleotides have been shown, by Schwartz, in our laboratory, to be attacked by ribonuclease and by venom phosphodiesterase.[52] Recently, we reported with Dr. Waehneldt and others that such polynucleotides, as well as calf

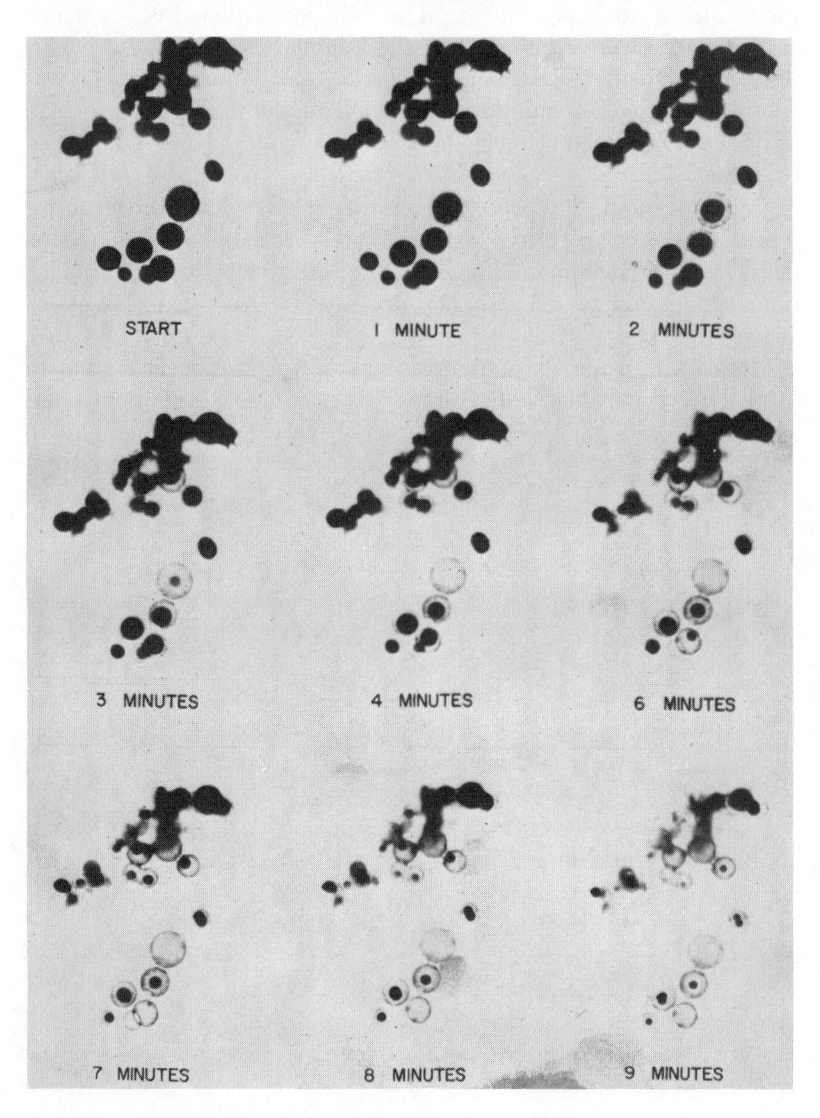

FIGURE 8. Photograph of pH effect followed by photograph of ultraviolet effect through quartz optics.

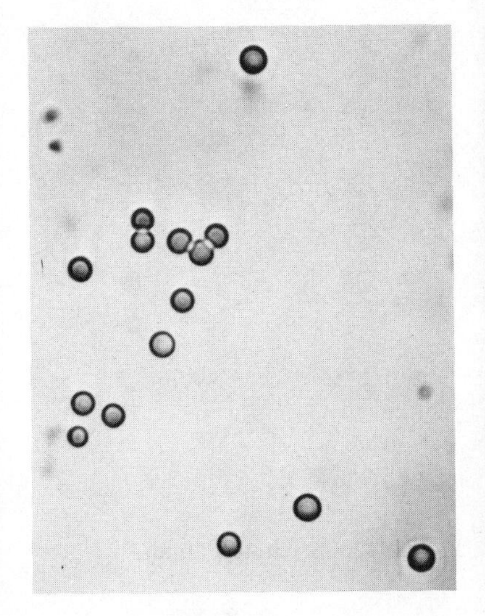

FIGURE 9. Microspheres composed of RNA and lysine-rich proteinoid.

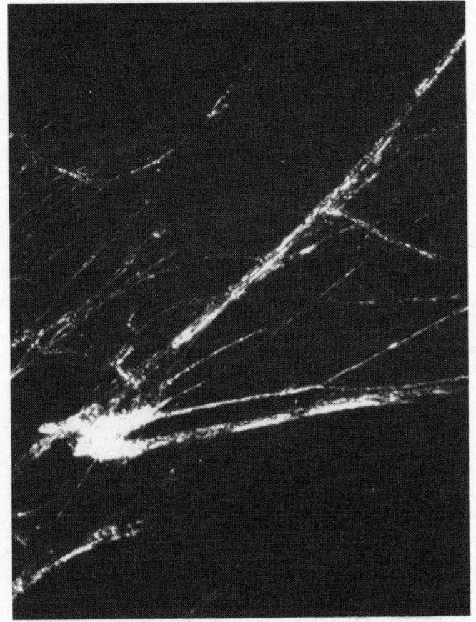

FIGURE 10. Photograph of fibrous complex of calf thymus DNA and of lysine-rich proteinoid.

thymus DNA and yeast RNA, bind with appropriate proteinoids to yield models of the various nucleoprotein particulates found in the contemporary cell, such as ribosomes, chromosomes, chromatin, etc. (Figure 9).

Figure 10 provides an example which shows fibers produced from lysine-rich proteinoid and calf thymus DNA. Very small microspheres result when RNA or thermal polyribonucleotides are used instead of DNA. The ratio of polynucleotide to basic proteinoid in such complexes tends to be quite constant. The fibrous and spherical morphologies are reminiscent of contemporary analogs as in chromosomes and ribosomes respectively. Those proteinoids that bind to form such particles have a ratio of basic amino acid to dicarboxylic amino acid above a minimum of 1.0.

With polyphosphoric acid and temperatures of 60–100°C, experiments in our laboratory have shown that either mononucleotides or amino acids might be polymerized.[52,18] Accordingly, these processes might ordinarily have occurred simultaneously. The suggestion of Calvin that proteins and nucleic acids might have arisen simultaneously in a primordial event is thus consistent with the experimental demonstrations.[53] Although model studies of prebiotic polynucleotides have been pursued, as indicated, a basic question that persists is that of how many properties models of primitive protein systems

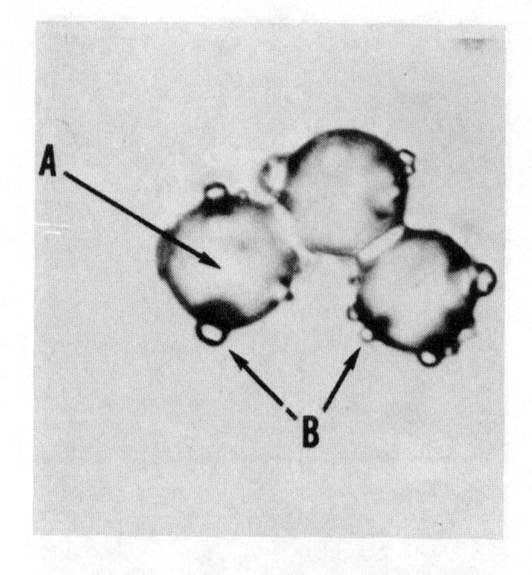

FIGURE 11. "Budded" proteinoid microsphere. A: microsphere; B: "bud."

FIGURE 12. Separated "buds."

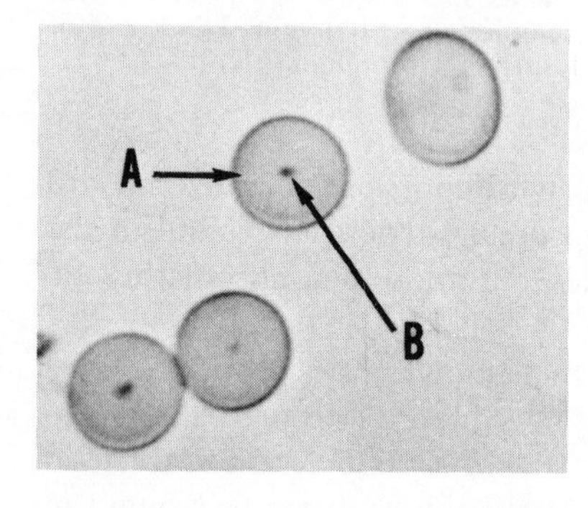

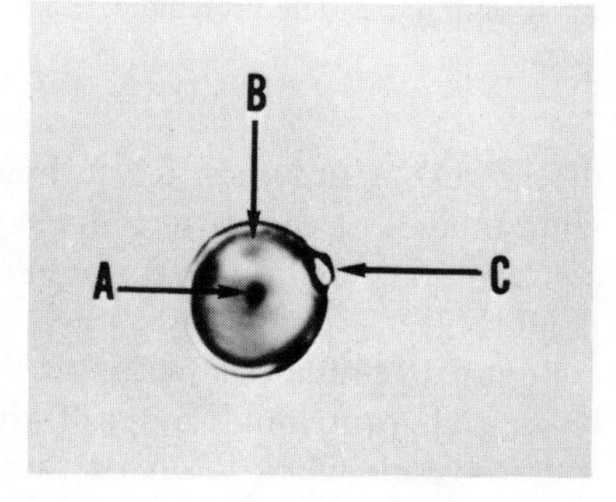

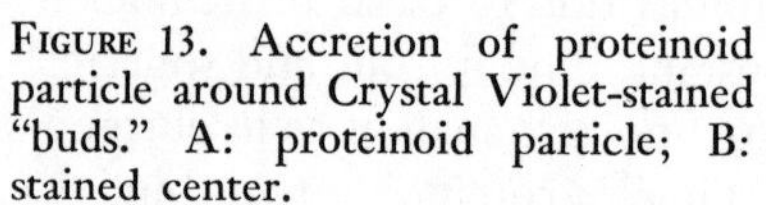

FIGURE 13. Accretion of proteinoid particle around Crystal Violet-stained "buds." A: proteinoid particle; B: stained center.

FIGURE 14. "Daughter bud" on microsphere. A: stained center; B: microsphere by accretion; C: "daughter bud."

might display *without* polynucleotides. This question especially deserves to be asked in view of the fact that proteinoids contain their own information and have sharply limited variation without any control by nucleic acids. Also, the self-assembling properties of proteinoid yield ultrastructure, double layers, fission, etc. without nucleic acid control.[7] Can then, for example, proteinoid microspheres multiply without polynucleotides present? In Figures 11–14 we observe how in a very simple manner proteinoid microspheres do, in fact, participate in the reproduction of their own likeness.[43]

In the first photomicrograph are seen a number of proteinoid microspheres which have been allowed to stand in their mother liquor for two weeks. On these are found "buds" which in appearance resemble buds on yeast. We first observed such buds in 1959. These "buds" can grow in size either while attached to the parent microsphere, or after separation. Removal can be accomplished at various stages of growth in size by shock—electrically, thermally, or mechanically. In the first two modes, we believe that some interfacial material is dissolved. "Buds" are seen in the second picture. In order to demonstrate the next event rigorously, the separated buds were stained with Crystal Violet. When the stained separated buds are allowed to stand in a solution of proteinoid saturated at 37°, and this is allowed to cool to 25°, the buds, in an appropriately sized vessel, "grow" by accretion to the size shown within one hour. The resultant units tend to be very uniform in size; the opposing forces are evidently precisely balanced under any given set of conditions.

In this manner we can visualize how proteinoid microspheres could simply first have developed the ability to participate cyclically in the reproduction of their own likeness. In manifesting such a process, a primitive organized structure would be functioning as a nearly complete heterotroph in that it obtained its large molecules by feeding on the environment instead of synthesizing them itself. Many theorists on the subject of abiogenesis have reasoned that the first organisms must have been heterotrophs. One may find the arguments in the writings of Oparin,[40] Haldane,[54] Pirie,[55] Pringle,[56] Horowitz,[57] Van Niel,[58] and others. To repeat, the hypothetical need for nucleic acid-mediated constraints on primitive protein are seen not to apply because of the internal constraints on the primary structure of proteinoid.[21]

Experiments in our laboratory have indicated how multiplication could also occur through the model of a primitive kind of binary fission[59] and growth by accretion, as well as through budding.

With cyclical proliferation due to budding or fission, as depicted, Darwinian selection can, conceptually, occur. Increasingly incisive experiments on Darwinian selection from nonbiological precursors have in fact been performed with acid proteinoid and neutral proteinoid.

The condensed description of this paper (in conjunction with other reports)[7] explains how a primitive organism capable of a kind of self-multiplication and possessing other salient properties could have emerged from primitive gases through the amino acids and subsequently through protein-like polymer. As perhaps need not be reemphasized, this model of a kind of primitive unit is clearly not a contemporary cell, at least not of the usual contemporary organism.

From this model, however, we visualize that in evolving to a contemporary organism a primitive self-replicating heterotroph would especially have had to develop an internal synthesis of protein and of polynucleotide.

The model processes which have been described are extremely simple. They consist of (a) heating above the boiling point of water, and (b) the intrusion of water. This simple sequence requires (a) geologically anhydrizing temperatures, e.g., those above 100°, and (b) sporadic rain or other common geological events of water such as drought or recession of the seas. These conditions (a) and (b) have been widespread geologically and, in fact, are quite widespread on the Earth today. The reactions are rugged, their occurrence is not easily disturbed by added substances since they are not solutes in aqueous solution, and the products have long-term stability. All details in the processes are found to be sequentially compatible.

To summarize, I use a 1966 quotation from Lederberg. "The point of faith is: make the polypeptide sequences at the right time in the right amounts and the organization will take care of itself. This is not far from suggesting that a cell will crystallize itself out of the soup when the right components are present."[42] The results described here (and others) show that when amino acids are simply and suitably heated, polypeptide sequences to at least some degree make themselves, and they do this in "the right amounts." The organization of the polypeptides does indeed take care of itself when water is added to thermal proteinoid, which appears to be the right component to crystallize out as a cell. Individual properties of proteins, polynucleotides, and of cells can be mimicked by other substances and units. Thermal proteinoid and its organized particles are the only chemically synthetic products, however, which have been shown to possess in *simultaneous* association and in lifelike inclusiveness properties of the contemporary cell and its structural polymer, with indications of the potential for further evolution to contemporary cells.

NOTES

1. Self Assembly of Structural Subunits. Symposium IV-4, Seventh International Congress of Biochemistry, Tokyo, Japan, 19–25 August 1967.

2. Ponnamperuma, C., Abstracts Seventh International Congress of Biochemistry, Tokyo, Japan, 19–25 August 1967, p. 483.

3. Fox, S. W., *Nature*, Lond., *201* (1964), 336.

4. Ponnamperuma, C., cited in *The origins of prebiological systems and of their molecular matrices,* edited by S. W. Fox (Academic Press, New York), 1965, 221.

5. Oro, J., cited in *The origins of prebiological systems and of their molecular matrices,* edited by S. W. Fox (Academic Press, New York), 1965, 137.

6. Ferris, J. P. & Orgel, L. E., *J. Am. Chem. Soc., 88* (1966), 3829.

7. Fox, S. W., *Nature,* Lond., *205* (1965), 328.

8. Akabori, S., Okawa, K., & Saito, M., *Bull. Chem. Soc. Japan, 29* (1956), 608.

9. Matthews, C. N. & Moser, R. E., *Nature, 215* (1967), 1230.

10. Fox, S. W., cited in *Evolving genes and proteins,* edited by V. L. Bryson and H. J. Vogel (Academic Press, New York), 1965, 359.

11. Hayakawa, T., Windsor, C. R., & Fox, S. W., *Arch. Biochem. Biophys., 118* (1967), 265.

12. Borsook, H. & Huffman, H. M., cited in *Chemistry of the amino acids and proteins,* edited by C. L. A. Schmidt (Charles C. Thomas, Springfield, Ill.), 1944, 822.

13. Dixon, M. & Webb, E. C., *Enzymes* (Academic Press, New York) 1957, 668.

14. Fox, S. W., *Amer. Scientist, 44* (1956), 347.

15. Steinman, G., *Arch. Biochem. Biophys., 119* (1967), 76.

16. Fox, S. W. & Harada, K., *J. Am. Chem. Soc., 82* (1960), 3745.

17. Vegotsky, A. & Fox, S. W., *Federation Proc., 18* (1959), 343.

18. Harada, K. & Fox, S. W., cited in *The origins of prebiological systems and of their molecular matrices,* edited by S. W. Fox (Academic Press, New York), 1965, 289.

19. Fox, S. W., Harada, K., Woods, K. R. & Windsor, C. R., *Arch. Biochem. Biophys., 102* (1963), 439.

20. Fox, S. W. & Harada, K., *Federation Proc., 22* (1963), 479.

21. Fox, S. W. & Nakashima, T., *Biochim. Biophys. Acta, 140* (1967), 155.

22. Block, R. J. & Keller, S., *Contrib. Boyce Thomson Inst. 20* (1960), 385.

23. Fox, S. W., *Nature* (1968).

24. Calvin, M., *Amer. Scientist, 44* (1956), 248.

25. Fox, S. W., *Amer. Naturalist, 87* (1953), 253.

26. Fox, S. W., Harada, K. & Rohlfing, D. L., cited in *Polyamino acids, polypeptides, and proteins,* edited by M. Stahmann (Univ. of Wisconsin Press, Madison), 1962, 47.

27. Noguchi, J. & Saito, T., cited in *Polyamino Acids, Polypeptides, and Proteins.*

28. Usdin, V. R. Mitz, M. A. & Killos, J. P., *Arch. Biochem. Biophys., 119.*

29. Rohlfing, D. L. & Fox, S. W., *Arch. Biochem. Biophys., 118* (1967), 122.

30. Rohlfing, D. L. & Fox, S. W., *Arch. Biochem. Biophys., 118* (1967), 127.

31. Fox, S. W. & Krampitz, G., *Nature, 203* (1964), 1362.

32. Fox, S. W., cited in *The origins of prebiological systems and of their molecular matrices,* edited by S. W. Fox (Academic Press, New York), 1965, 361.

33. Durant, D. & Fox, S. W., *Federation Proc.*

34. Oshima, T., *Federation Proc., 26* (1967), 451.

35. Krampitz, G. & Hardebeck, H., *Naturwiss., 53* (1966), 81.

36. Rohlfing, D. L., *Arch. Biochem. Biophys., 118* (1967), 468.

37. Krampitz, G., Diehl, S. & Nakashima, T., *Naturwiss.* (1968).

38. Needham, A. E., *The uniqueness of biological materials* (Pergamon Press, Oxford), 1965.

39. Alcock, R. S., *Physiol. Revs., 16* (1936), 3.

40. Oparin, A. I., *The Origin of Life on the Earth*, 3rd ed. (Oliver and Boyd, Edinburgh), 1957.

41. Wald, G., *Sci. American, 192* (Aug. 1954), 44.

42. Lederberg, J., cited in *Current topics in developmental biology, 1*, edited by A. D. Moscona and A. Monroy (Academic Press, New York), 1966, ix.

43. Fox, S. W., McCauley, R. J. & Wood, A., *Comp. Biochem. Physiol., 20* (1967), 773.

44. Smith, A. E. & Bellware, F. T., *Science, 152* (1967), 362.

45. Fox, S. W. & Yuyama, S., *Ann. N.Y. Acad. Sci., 108* (1963), 487.

46. Fox, S. W. & Yuyama, S., *J. Bacteriol., 85* (1963), 279.

47. Murray, R. G. E., cited in *The Bacteria*, I, edited by I. C. Gunsalus and R. Y. Stanier (Academic Press, New York), 1960, 55.

48. Wetlaufer, D. B., *Advances Protein Chem., 17* (1962), 303.

49. Fox, S. W., McCauley, R. J., Fukushima, T., Windsor, C. R. & Montgomery, P. O'B., *Federation Proc., 26* (1967), 749.

50. Waehneldt, T. & Fox, S. W., *Biochim. Biophys. Acta, 134* (1967), 1.

51. Ponnamperuma, C., cited in *The origins of prebiological systems and of their molecular matrices*, edited by S. W. Fox (Academic Press, New York), 1965, 234.

52. Schwartz, A. & Fox, S. W., *Biochim. Biophys. Acta, 134* (1967), 9.

53. Calvin, M., *Bull. Am. Inst. Biol. Sci., 12* (Oct. 1962), 29.

54. Haldane, J. B. S., *New Biol., 16* (1953), 12.

55. Pirie, N. W., *New Biol., 16* (1954), 28.

56. Pringle, J. W. S., *New Biol., 16* (1954), 54.

57. Horowitz, N. H., *Proc. Natl Acad. Sci., 31* (1945), 153.

58. Van Niel, C. B., cited in *The microbe's contribution to biology*, edited by A. J. Kluyver and C. B. Van Niel (Academic Press, New York), 1956, 155.

59. Fox, S. W., & Yuyama, S., *Comp. Biochem. Physiol., 11* (1964), 317.

Evolutionary Parsimony

Ralph E. Thorson

Modern biologists, particularly teachers, in order to impress on their students the glamor of the field, overemphasize the diversity of life, both plant and animal. For instance, a single celled amoeba is certainly different in outward form from an elephant; likewise, an algal cell differs from a mammoth redwood. The purpose of this discussion, however, is to emphasize the converse picture. Evolution has produced diverse forms, but it is extremely apparent that evolution also displays what we choose to label evolutionary parsimony. As one surveys living matter, it is apparent that there is a conservatism present in all levels of biological organization, molecular, cellular, organismic, population, and community. Countless examples of parsimony are evident and the examples chosen in this discussion are not meant to be all-inclusive, but are to serve as a blazed trail for those who accept our idea.

To the comparative physiologist and molecular biologist this parsimony is probably more apparent than to other biologists. However, it is not infrequent that the molecular biologist analyzing the amino acid sequence in proteins overlooks the forest for the trees. The writer, trained primarily as a zoologist, will probably emphasize animal forms, but certain obvious parsimonious factors are apparent in such diverse phylogenetic groups as the Kingdoms Protista, Plantae, and Animalia.

A. MOLECULAR LEVEL

Laboratory and theoretical studies on prebiochemical and biochemical evolution have in a sense dictated the conservative niggardliness we characterize as parsimony. In the early atmosphere and on the earth, there were a finite number of elements and compounds available for biochemical evolution, a finite number of stimuli for compound formation, and a finite number of complex molecules produced. Today, the possible concatenations of amino acids which can make up a protein molecule appear staggering to us. (In a protein with 100 amino acid residues with 20 plus amino acids available, if chance determined each amino acid residue, 100^{20} proteins would be possible.)

Nonetheless, as more and more detailed amino acid sequences become known it is apparent that the varieties and numbers of combinations or sequences developed is not that extraordinary. There is a basic conservatism in the structure of proteins. Useful examples are cited by Anfinsen (1967). The elucidation of the amino acid sequences of trypsinogen and chymotrypsinogen indicates that a large part of the sequence is common to both (Fig. 1).

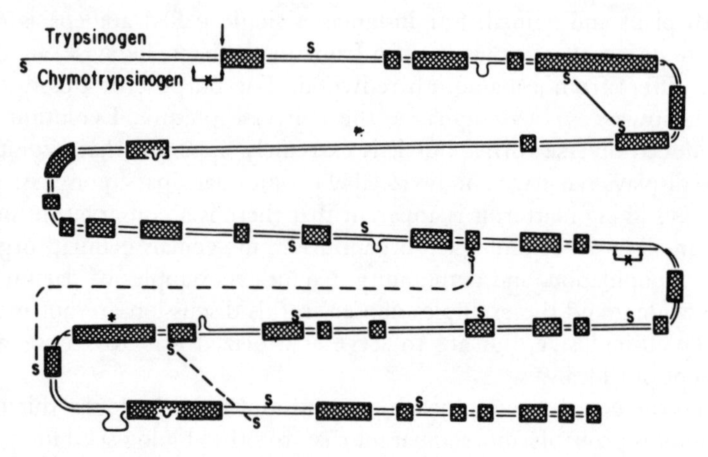

FIGURE 1. Sequence homologies in trypsinogen and chymotrypsinogen.

This similarity extends more particularly to the (1) sections involved in the catalytic activity of the compounds, (2) in two of the disulfide bonds which help determine secondary and tertiary structure, and (3) also in the activation site at which peptide bond cleavage leads to activation of the zymogens. Anfinsen concludes that it is "extremely likely that these proteins are derived from the same primitive precursor." In an extremely interesting paper, Epstein (1967) by a careful study of amino acids suggests that natural selection has acted in the evolution of homologous proteins to favor amino acid substitutions that would be "compatible with the retention of the original conformation of the protein and conversely to eliminate those that would not." Steinman (1967) likewise points out that in contradistinction to the suggestion that peptide synthesis occurring nonenzymically on the primitive earth would be a totally random event, evidence can be deveolped that the interaction of any one given amino acid with another follows well-defined statistics based on the relative reactivity of each amino acid.

Cytochrome c has been found in a wide variety of organisms. These chromoproteins from species as diverse as man, chicken, pig, horse, and ox representing mammals, chicken representing birds, tuna representing the fishes and yeast, the fungi, display striking similarities in amino acid sequence (Fig. 2).

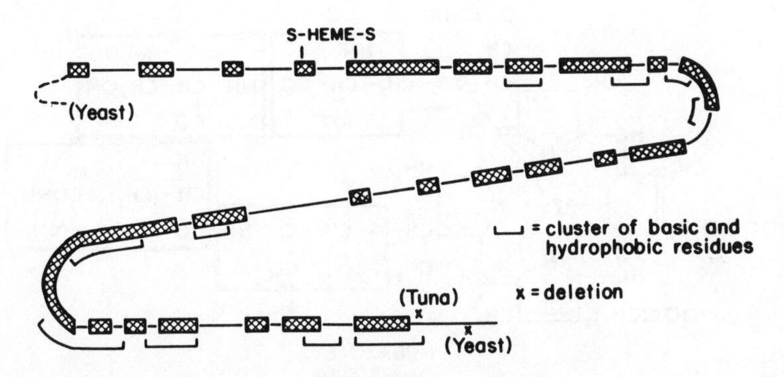

FIGURE 2. Areas of identity in the sequences of cytochromes *c* (from man, chicken, pig, horse, ox, tuna, and yeast).

Clusters of hydrophobic and basic amino acid residues are present as fairly regular features of all cytochromes (Anfinsen, 1967; Jukes, 1966). Jukes has stated that "evolutionary changes in the cytochromes c are evidently extremely slow." This statement might more properly be phrased as "evolutionary changes in the cytochromes c are evidently parsimonious." If viewed from a conservative viewpoint, Juke's work "Molecules and Evolution" is an extremely cogent argument for our thesis of evolutionary parsimony. There are countless examples of this thesis at the molecular level and each day sees publication of additional evidence.

The importance of histidine residues and their spatial relationship with respect to activity in enzymes is striking. The recent demonstraton by Liu (1967) that the position of these residues with respect to each other determines the specificity of streptococcal proteinase is a basic finding. Hoffee and others (1967) also point out the functional significance of histidine residues in rabbit muscle aldolase. The functional role of histidine is not restricted to ribo-nucleases or aldolase but is likewise important in trypsin and chymotrypsin and a rather large series of recently described structure-activity relationships. Another feature which bears scrutiny is that in these studies and those on DNA generally, tertiary structure takes the form of an α-helix. Spatial geometry seems to dictate a recurrence of this space-saving economy. A solid shape, if it could be imagined, for these molecules restricts surface area but complete linearity would not conserve space.

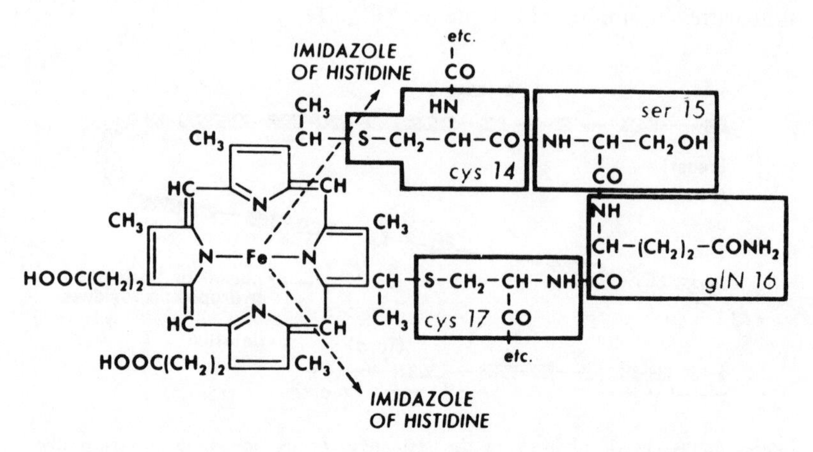

FIGURE 3. The heme group and its attachment to human cytochrome *c*.

One of the prime arrangements for determining spatial geometry in proteins is the disulfide bridge which is found universally in proteins studied even occurring in a proteinoid compound as small as insulin (with two disulfide bridges between the A and B chains of amino acids). Evolutionary economy apparently dictated that few experiments to determine shape with such a configuration were possible.

Energy relationships and the coinage of energy in living things were likewise not subjects of great evolutionary experimentation. Adenosine monophosphate (AMP), adenosine diphosphate (ADP) and adenosine triphosphate (ATP) are universal in the plant and animal kingdoms. Thus in the unfolding of energy relationships at least the terminal agents of energy transformation give evidence for evolutionary parsimony.

When one considers two basic energy processes in living organism –photosynthesis and oxidation–the synthesis and breakdown of energy rich materials, evolution again has been most conservative. Such substances as chlorophyll in plants, hemoglobin in higher animals, hemocyanin in lower animals, and cytochromes of plants and animals all are tetrapyrrholes or protoporphyrins conjugated to proteins and differ primarily on the structure of the associated protein. The protoporphyrins are very similar and differ from each other primarily in possessing a "nucleus" of magnesium in cholorophyll, iron in hemoglobins and cytochromes and copper in hemocyanin. Interestingly, in relation to our discussion of enzymes, it appears that the functional porphyrins are connected to the protein part of the molecule at the imidazole nitrogen of histidine.

B. CELLULAR LEVEL

The recent plethora of superb electron photomicrographs suggest that evolution has not experimented widely with morphology. Cell structure has what might be called a basic monotony. In forms possessing a nucleus, there is little to differentiate on the basis of this structure either plants or animals. The cytoplasmic structures reveal a remarkable sameness–the mitochondria, Golgi apparatus, cilia or flagella, ribosomes, endoplasmic reticulum, even the cell membranes are similar in all animals and plants. The centrioles, present in animals but absent in plants, are an intriguing exception. This identity of structure illustrates a principle mentioned earlier –an economy of packaging is achieved by increasing surface areas in a semiconservative way. An increase in area provides for increased areas of physical and chemical activity yet the conservation principle operates to keep the package small. An obvious example is the interdigitation of the cristae of mitochondria; also the lining up of ribosomes along the endoplasmic reticulum, and the lamellation in chloroplasts.

Other obvious similarities include the basic structure of centrioles, cilia and flagella. Likewise, observe the remarkable likeness of primitive chloroplasts–which will eventually deal with organic synthesis –and mitochondria, which deal with production of energy for the cells. The chloroplast produces fuel from the energy of sunlight, water and carbon dioxide and the mitochondria produce from this sugar, carbon dioxide, water and energy.

Depending on cellular function, it is possible to see differences in cellular morphology which are a reflection of the activity of the cell; i.e., the rough endoplasmic of a rat liver cell (the roughness due to the number of ribosomes on the reticulum) vs. the smooth endoplasmic reticulum of the interstitial cell of the testis of an opossum.

All in all, the similar morphology of all cells which reflects the common activities of those cells is a cogent argument for evolutionary parsimony.

C. ORGANISMIC LEVEL

Examples of evolutionary parsimony at the organismic level are legion. Again, this survey is not expected to cover or anticipate all examples and I must be excused if many of the examples emerge from the field of parasitology, the area of my research interest.

It is apparent that the evolutionary process is not going to spend money in the form of energy if it is unnecessary. The most remarkable instance of this expression is the loss of a gut in the Cestoda, the tapeworms. In the phylum Platyhelminthes, the freeliving Turbellaria (planaria) have a saclike intestine, as do the parasitic Trematoda, but the tapeworms all lack a gut. It is interesting that these parasites as adults almost without exception are dwellers in the intestine of their hosts. This location is ideal for these parasites in that there is no need for a gut of their own for processing food which is ingested. The food is already in an assimilable form. The energy required in embryogeny for gut formation is thus conserved.

Parasites are generally considered as degenerate because of the loss of so many structures. I submit that this is merely an example of evolutionary parsimony and in no way implies degeneracy.

As more and more research is done with hormones, it becomes apparent that substances with similar physiological activities are present in vertebrates and invertebrates (Williams, 1961; Schneiderman & Gilbert, 1964). Recent work with invertebrates has delineated chemicals which are involved in growth and maturation particularly of arthropods–insects and crustaceans. There is one citation of substances with juvenile hormonelike action in the protozoan *Nosema* (Fisher, 1963). Our own work (Thorson and others, 1968) with cestodes, trematodes, and nematodes indicates that similar chemicals play a role in growth and maturation of these parasites. When one considers the complexity of the life cycles of parasites generally, mechanisms to control moulting, maturation in the many and various life cycle stages would seem essential. That they seem to be similar to the mechanisms investigated in the much more highly evolved arthropods and even vertebrates is yet another instance of evolutionary economy. Even more remarkable, however, is the fact that these substances are also found in plants (Slama and Williams, 1965). What function these substances serve in plants or whether they are simply intermediate in a lengthy synthetic cycle is not clear, but that the synthetic cycles for terpenes and subsequently, sterols are similar if not identical in plants further illustrates the parsimony thesis.

Our work on the behavior of parasitic helminths (McCue and Thorson, 1964, 1965) in a thermal gradient is another example. It appears that parasitic helminths (nematodes, trematodes, cestodes, and acanthocephalans) have developed an exquisitely sensitive mechanism for detecting temperature differences (conservatively estimated at 0.07°C but probably much more sensitive). These parasites always migrate up the thermal gradient from lower to higher temperatures. The conservative interpretation of this behavioral aspect is that the parasites will choose a temperature optimum. However, this is not the case (Fig. 4). Rather, the worms continue upwards in the gradient until they are killed by excess heat. The normal homeostatic mechanism would dictate that a negative feedback mechanism would make the parasites oscillate back and forth around a temperature which would be optimal. They do not. Nonparasitic helminths such as *Enchytraeus* (a fresh water oligochete)

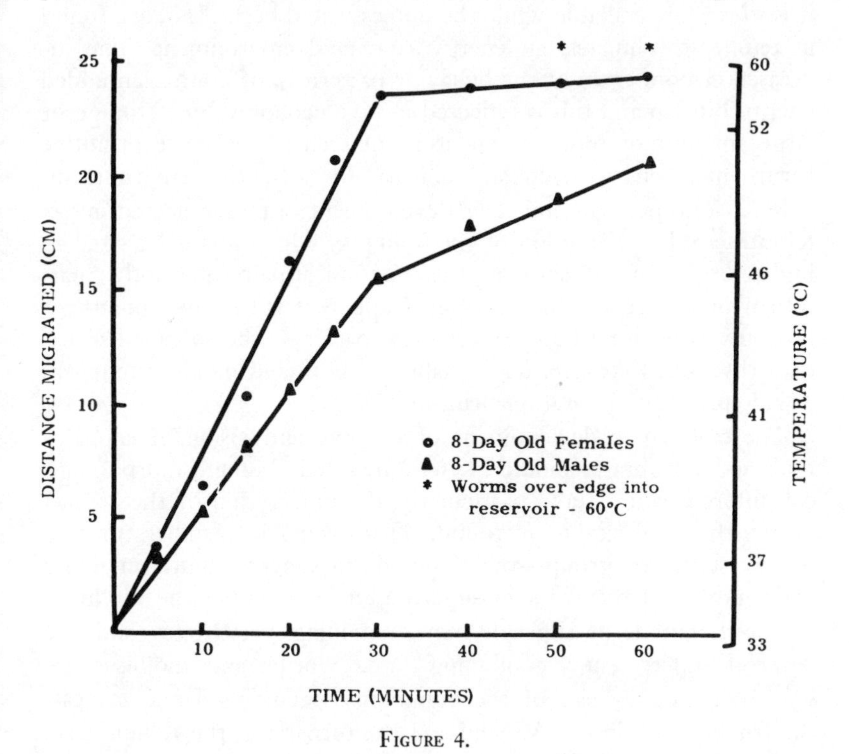

FIGURE 4.

and *Dugesia* (a planarian) do not show this response. The closed loop nervous system is not present. Why? The organisms have evolved a parasitic existence which makes an energy expenditure for negative feedback an extravagance. The host, on whom the parasite depends, will (if it is to be successful), insure that a lethal temperature will not be met. The sensitivity of the temperature recognition, however, is economical in that if a host is in the vicinity it will be recognized. But the necessity for a shut-off mechanism on this behavioral pattern is negated by the fact that the same host will preserve the parasite from lethal high temperatures.

The classical studies of bee language by Von Frisch (1953) and more recently by Esch & Esch (1965) illustrate yet another facet of evolutionary economy in behavioral mechanisms. The basic communication signal in primitive forms is scent and sound while in the more highly evolved and domesticated bees, this basic signal is augmented by elaborate motions called "dances" which direct the bees to a food site. The more primitive forms exist in an energy rich environment (Brazilian tropical rain forests) and need little in the way of directional and distance signals to obtain food because it is so readily available while the domesticated bees, *Apis*, are found in temperate climates, an energy restricted environment. The increased elaborateness of the behavior pattern is, of course, an added energy burden and this is reflected in hive economy by a storage of large amounts of food, a regulator not seen in the more primitive forms in an energy rich and climatically attractive environment.

Behavioral patterns and their development in these selected invertebrates are but a fraction of the examples which might be cited in higher organisms—all showing the economical pattern. Another pattern which more and more becomes apparent is the rhythmicity of not only behavioral but physiological patterns: the so-called circadian rhythms. Likewise, territoriality is more and more common as a widespread behavioral mechanism.

The embryonic development of the eye and vision in cephalopods and vertebrates should likewise be cited. Though morphological differences in detail (particularly the innervation of the retina) occur, the similarity is profound. This occurs in members of two widely separated groups—the Echinoderm superphylum containing vertebrates and the Annelid superphylum containing the molluscs.

The Echinoderm superphylum containing vertebrates and the Annelid superphylum containing some worms and molluscs are separated on the basis of the commonly occurring larval stages—the trochophore in the Annelid and the tornaria in the Echinoderm

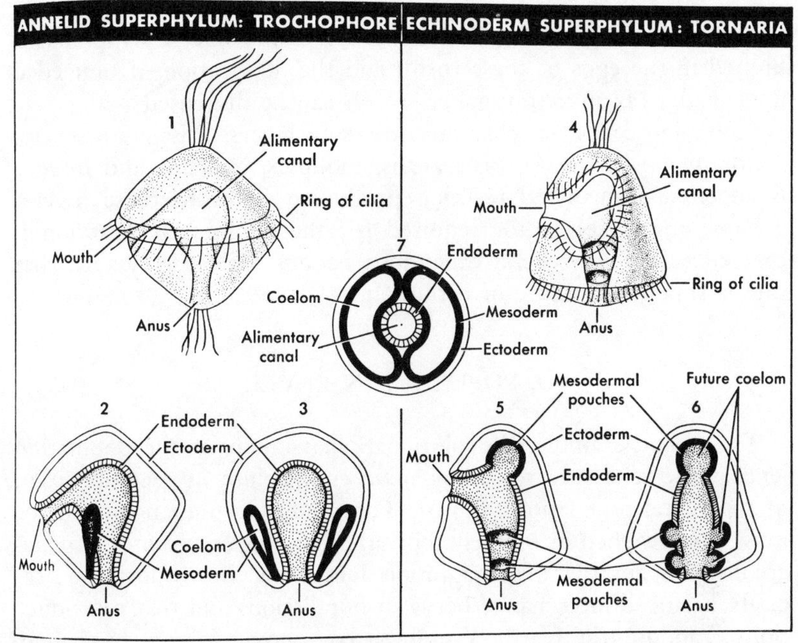

FIGURE 5.

superphyla (Fig. 5); morphologically, these two larval stages are not too dissimilar. The position of the ring of cilia around the oral opening in the tornaria and subequatorially in the trochophore constitute the most obvious difference. The development of mesoderm, however, is quite different–schizocoelous in the annelid group and the enterocoelous in the echinoderms. The basic morphology, however, is so similar as to suggest an evolutionary advantage and economy of this general shape.

With our more sophisticated understanding of the functioning of DNA, the process of chromatin diminution in certain of the Aschelminthes, particularly the nematodes likewise illustrates conservatism. The observation that at various and specific times during cleavage, fractions of the chromosomes in certain somatic cells are discarded and do not migrate during anaphase to the new daughter cells' nuclei indicates in retrospect that the function of this piece of coded information is no longer needed for development. Interestingly, these morphological observations were made by Boveri in 1893. One should, however, point out what seems to be an evolutionary disadvantage in the conservatism of chromatin diminution. Except for the germinal cells producing eggs and sperm, which do not undergo chromatin diminution, the other cells lose the capability of dedifferentiation and thus repair of damaged tissues is not possible. However, the store of energy material available for embryonic development is limited in the eggs of these forms and the duplication of unneeded nucleic acids is an extravagance which can be dispensed with.

Bioluminescence, a characteristic which has arisen in several major groups, bacteria, protozoans, molluscs, annelids and insects, is an enzyme-mediated reaction producing light. In these diverse groups, one major factor required for the luminescent reaction is oxygen (although several exceptions occur) indicating yet further the parsimonious nature of evolution.

D. POPULATION LEVEL

The classic example of evolutionary parsimony at the population level is the tendency toward genetic equilibrium in recombination of biparental populations. Given: That mutation must not occur or must have reached its own equilibrium, that chance changes in genes are insignificant, that gene migration does not occur or must be a precisely balanced interchange between populations and that reproduction is random, the Hardy-Weinberg law conserves genetic change.

E. COMMUNITY LEVEL

At the community level of biological organization, it is somewhat more difficult to develop an example of our principle. Since a community is composed of an aggregation of populations each subscribing to the parsimony axiom, one might expect them in concert to produce a structure in communities which is coherently recognizable. The best example, on the structural level, is possibly the most severe, the deserts of the world. As described by Odum (1963), the annuals avoid drought by growing only when there is adequate moisture. The shrubs of desert with numerous branches arise from a short basal trunk and have small thick leaves that may be shed during dry periods; their reason for surviving is an ability to anticipate drought and thus to become dormant before they wilt. The succulents, cacti in the New World or the euphorbias of the Old World store water in their tissues. The microflora such as mosses, lichens, and blue green algae remain dormant in the soil but respond quickly to cool or wet periods. Some animals such as reptiles and some insects, adapted by reason of impervious integuments and dry excretions, survive well in desert communities by utilizing the water produced in carbohydrate breakdown. Mammals are generally not successful in deserts, but secondary adaptation has selected some good desert dwellers. For example, some rodents particularly have adapted to a desert existence by passing concentrated urine and being active only at night. Most important, however, is that anyone who sees a desert community whether Old or New World recognizes it as a desert, because the species present, though taxonomically widely different in the two areas, have evolved the same economical devices to enable survival in the barren areas.

F. CONCLUSION

In this brief and very selective review, we have attempted to set forth evidence for a principle of adaptation which we call evolutionary parsimony. Hopefully, others can select and cite examples far better than those we have chosen.

NOTES

Anfinsen, C. B. 1967. Molecular structure and the function of proteins. In *Molecular Organization and Biologic Function*, Ed. J. M. Allen, Harper and Row, New York.

Boveri, T. 1893. Ueber die Entstehung des Gegensatzes zwischen den Geschlechtzellen und den somatischen Zellen bei *Ascaris megalocephala,* nebst Bemerkungen zur Entwickelungsgeschichte des Nematoden. Sitsungsb. Gesellsch. *Morphol. u. Physiol. Munchen 8*:114–125.

Epstein, C. J. 1967. Non-randomness of amino acid changes in the evolution of homologous proteins. *Nature 215*:356–359.

Esch, H. and I. Esch. 1965. Sound: an element common to communication of stingless bees and to dances of the honey bee. *Science 149*:320–321.

Fisher, F. M. Jr. 1963. Production of host endocrine substances by parasites. *Ann. N.Y. Acad. Sci. 113*:63–73.

Frisch, K. von. 1967. *The Dance Language and Orientation of the Bees* Harvard University Press, Cambridge, Mass.

Hoffee, P., C. Y. Lai, E. L. Pugh, and B. L. Horecker. 1967. The function of histidine residues in rabbit muscle aldolase. *Proc. U.S. Nat. Acad. Sci. 57*:107–113.

Jukes, T. H. 1966. *Molecules and Evolution*. Columbia University Press, New York.

Liu, T. 1967. Demonstration of the presence of a histidine residue at the active site of streptococcal proteinase. *J. Biol. Chem. 242*:4029–4032.

McCue, J. F. and R. E. Thorson. 1964. Behavior of parasitic stages of helminths in a thermal gradient. *J. Parasit. 50*:67–71.

McCue, J. F. and R. E. Thorson. 1965. Host effects on the migration of *Nippostrongylus brasiliensis* in a thermal gradient. *J. Parasit. 51*:414–417.

Odum, E. 1963. *Ecology*. Holt, Rinehart and Winston, Inc. New York.

Schneiderman, H. and L. Gilbert. 1964. Control of growth and development in insects. *Science 143*:325–333.

Slama, K. and C. M. Williams. 1965. Juvenile hormone activity for the bug *Pyrrhocoris apterus*. *Proc. U.S. Nat. Acad. Sci. 54*:411–414.

Steinman, G. 1967. Sequence generation in prebiological peptide synthesis. *Arch. Biochem. and Biophys. 121*:533–539.

Thorson, R. E., G. Digenis, A. Berntzen, and A. Konyalian. 1968. Biological activities of various lipid fractions from *Echinococcus granulosus* scolices on in vitro cultures of *Hymenolepis diminuta*. *J. Parasit. 54*:970–973.

Williams, C. M. 1961. Present status of the juvenile hormone. *Science 133*:1370.

Some Uncertainties of Evolution

Pierre P. Grassé

In the quite informal comment to follow I shall focus my attention on certain problems of evolution. Evolution is no longer a hypothesis. Today we can think only in evolutionary terms.

Why? Because of an immediate practical reason; namely, that we have constant confirmations, and also of reasons of a methodological and philosophical order. Is the living world intelligible without the notion of evolution? No; without this concept we could no longer understand anything. We would have to resign ourselves to understanding neither the vegetable forms, nor the animal forms, nor the human form. I shall try to show you, through examples, how the concept of evolution is verified.

A bird's-eye view of evolution, from the oldest fossil age to our day, shows us that evolution did not take place haphazardly. It is not chaotic or illogical. The forms succeed each other in a certain order, though no doubt with accidents, backslidings, meanderings, sometimes catastrophies. But when we consider the spectacle from on high we perceive that evolution has led above all else towards the complication of forms. This complication has occurred in ways that are not numerous. Take into account that the animal world can be reduced to about twenty principal plans of organization—perhaps twenty-four or twenty-five, but in reality nearer twenty.

At the same time we perceive an extremely important phenomenon, which Teilhard de Chardin brought to our attention; namely, that in certain respects this organic complication was accompanied by a loss of functions. This does not signify, as some people have believed, that evolution in this case was regressive. Not at all. Because man does not have the same functions as the unicellular algae of the Precambrian period, we cannot conclude that he has had a regressive evolution. If we admitted this notion of regressive evolution we would imply that the human brain is not an important acquisition.

Yet this brain, as I will show you shortly, has completely modified the biological aspect of our universe.

At the same time we see that living animal beings liberate themselves increasingly from their environment because they gain in psychical power. In other words, their behavior becomes more complicated and detaches them from the environment, gives them a certain liberty or independence. They are not automatically subject to the laws of their environment, thanks to certain of their reactions.

This psychical power kept on increasing and attained at the end of the Tertiary, in full Quaternary, the development of the human brain we know, the brain of *Homo sapiens*.

This great fresco, therefore, has an orientation. Evolution was not brought about by chance. There are laws of evolution which we know, but they are superficial laws. The deeper and ultimate laws remain unknown.

Biology is full of unknowns and this is not surprising. The physicist comprehends relatively simple phenomena, in fact very simple phenomena, for which the number of parameters is limited. In biology, on the other hand, the number of parameters is extremely high. I would even say that there are so many that, in general, we are unable to determine them exactly. We can imagine that a given phenomenon depends on two, three, or four causes, but the analysis may show us that in reality there are many more causes. I do not know any biological phenomenon of which we can actually say that we master all the causes.

Consequently, when we speak of biological determinism our language lacks rigor and can only be approximate. This means that biological phenomena are very difficult to study since we do not know the laws governing living beings. I do not say "living matter," for we do not know what "living matter" is; we have never seen it. We only know living beings, namely organisms that have a unity, an individuality. These organisms are made of a certain matter, but we never know this matter in an isolated state. This is one of the great difficulties in explaining the appearance of life on earth. It is the passage from the unorganized to the organized, and this "organized" is not simply organized like a crystal, but is a heterogeneity of which the parts present an absolutely obligatory architecture. If this architecture were not there, the machine would not work.

A living being with an increasing psychical power obtains increasing independence as regards the surrounding medium. This is, therefore, one of the rules of evolution in the animal realm, and we must ask ourselves if there are not other means of verifying the hypothesis given by paleontology with the help of comparative anatomy.

We see a certain number of relatively simple organisms, not only the unicellular but also the cellular which have only two germ layers, and so are called "diploblastic." These organisms, according to theory, and in particular the theory of evolution, are placed at the very base of the animal realm. But we could not know this until a few years ago since we had not found them fossilized in good condition in the oldest layers. According to prevision these organisms should be very distant and very primitive. This prevision has been confirmed during the past few years.

Research was conducted in very ancient, Precambrian formations—perhaps five or six million years earlier than the Cambrian—and so pushes our findings back to nearly a billion years. What was there at that epoch? There were the "diploblastic forms" which are found, perfectly preserved, in deposits in South Australia, in the region of Adelaide, in the hills of Ediacara. A quantity of medusae were found, fixed forms that were probably also diploblastic, sponges, and —what is also interesting—some forms that forecast worms. The triploblastic state had probably already been reached. This is a magnificent confirmation of previsions, but there are other confirmations.

Let us ask for proofs from the most modern techniques of biology. The knowledge we now have of the cellular ultrastructure enables us to affirm something which was known in general but which we could not talk about because we were not familiar with all the details. This "something" is the unity of the cellular plane.

In the whole animal plane, whether it be that of the protozoa or a unicellular alga, or the cell of a horse, or the cell of a man, or the cell of a plant, not only is the plan the same but the tools that intervene in the physiology, the metabolism of the cell are identical. The "kitchen battery" is everywhere the same. We always find the nucleus, with its chromosomes composed of the same elements, with the nucleoli, with the same mitochondria, with the same Golgi bodies, the same ergastoplasm. All the elements are present. The plants have one more element—the chloroplasts in which photosynthesis takes place. These structures store the energy of light in chemical bonds and permit other living creatures to survive.

This can only be explained and understood by assuming a unity of origin. The further we advance the sounder becomes the monist theory—the theory which only fifty years ago made hair stand on the heads of certain philosophers. There is now no doubt that all living beings derive from the same living matter.

This does not mean that at the beginning there was only one matter, one living being. There were perhaps many, but there was only one living being which survived and all of us are its descendants. All actual nature arose from this same matter, this same organism. But there are other arguments to show the veracity of the evolutionist theory.

I.

All multicellular animals, even the simplest, have incontestably the same ancestor. The proof is given by the structure of their reproductive elements. All have the same structure, the same plan of organization. The egg likewise shows the same properties: fecundation occurs following the same process, be it the egg of a sea urchin or the egg of a mammal. The diastase, the ferments, the enzymes which intervene are in the same place. All spermatozoa possess the same properties.

The changes occur in the genes, namely the elements which transmit the hereditary characteristics. But for all the rest, it is the same in all living beings: the big processes, the broad features, remain the same everywhere.

II.

I could multiply for you the examples taken from paleontology, zoology or parasitology, but this would be useless. This much having been said, and having affirmed my deep conviction in the justification of evolution (it is a kind of credo, firmly anchored in me), we are now free to see the difficulties when we try to explain what we run up against.

We speak, for instance, of problems of adaptation. For a long time adaptation was held in very poor repute. Chauvin spoke of it in connection with Rabot. I was a pupil of Rabot and we all know what this anti-Bernardin de St. Pierre said. But it is very dangerous to be "anti-something." This does not prevent my thinking that Rabot played an important part in the story all the same, and that he was an outstandingly intelligent man. His mind was more critical than constructive but he was really very gifted.

He was not angry when I wrote more than thirty years ago in a treatise on the action of parasites and parasitism that evidently legs are not very useful. A legless cripple lives quite nicely and is able to procreate. Nevertheless I said that it was better to have two legs. Rabot sensed the criticism, but he took it very well.

Adaptation is a fact that is evident and we need not bring final causes into consideration as Voltaire does in the *Philosophical Dictionary* discussing the word "tear." He pities those unfortunates who are not clear-sighted enough to understand that the eye is made to see and the ear is made to hear. Certainly we do not have the right to say "for the purpose of" and so we will simply state that the eye sees and the ear hears, which does not change the problem very much.

Chauvin told you that the antifinalist Rabot did not like to speak of the eye. Neither did Darwin. He wrote to his friend and colleague, the botanist Asagré: "When I think of the eye, I develop a fever." Why? Because he recognized his inability to explain it.

Julian A. Huxley gave a lecture in the Palais de la Découverte: in 1946. He is a convinced neo-Darwinian, indeed a hyper-super-neo-Darwinian. But not all who call themselves so are like that. For instance, J. B. S. Haldane, the great theoretician of neo-Darwinism, once wrote to me: "It was evidently necessary that mutations should not be aleatories." He wrote me this when he was in Bombay, after he had become converted to Buddhism.

Nowadays, on the other hand, we see Dobzhansky, a great exponent of neo-Darwinism, who, it would seem, takes the criticisms we address to him so much to heart that he has asked me to write a foreword for a simon-pure neo-Darwinian book.

The eye presents very difficult problems, indeed. The systems where there are co-adaptations between the parts, the systems where the organ does not function if there is not precise adjustment, raise unanswerable questions. It is the existence of the co-adaptations between the parts which makes it so difficult to understand a relatively simple organ like the eye. This organ had to be complicated from the start or else it could not function.

In studying the invertebrates we can follow to a certain degree the way things come into being. At present there is a pupil of our friend Drach who is writing a thesis on the ultrastructure of the eyes of the polychaetous annelids. I am her scientific advisor. One sees how, starting from simple elements, one can progress to ultrastructures. But it is a question of very simple forms and not truly of the eye, the cameralike eye, the eye with the chamber of the vertebrates. That eye demands extraordinary precision.

Chauvin told you that the eye has deficiencies. It has been demonstrated that these deficiencies are qualities. It is because the eye presents certain chromatic aberrations, certain aberrations of sphericity, that it can of itself correct errors. There are considerable advantages in all of this.

The eye deserves to be treated as a true marvel, for it is an astonishing and extraordinary instrument. Think of the difficulties propounded by the organization of its parts! It is necessary for these parts to have appeared in a certain order, for if they do not appear in that order the system does not function. Imagine that there were no connection between the crystalline lens and the retina. It would no longer work at all. But as a matter of fact this connection exists because the brain in its anterior has a zone which organizes the retina and the retinal cupula is an organizer for the crystalline lens. Everything holds together. It is not a system of independent variables; the variables must be coordinated. Almost always we find this coordination, for uncoordinated variables produce nothing.

When we consider the ear, matters grow even worse. The organ of Corti had to evolve in a totally and strictly coordinated manner or else the ear would not function.

But until now we have taken into consideration only a very small part of the difficulties to be resolved. The eye must put itself into accord with the brain, and at that moment it must enter into relation with the centers. These centers must analyze the picture which must then be sent into "organizing zones"—sensorial zones which are to be found in the cerebral cortex. We have here at least three stages: the receptor, the transmitter with analyzer, and finally the part which gives the sensation with integration. It is an extraordinarily complicated operation.

First, there has to be a system of variations in the interior of the eye. Then two mechanisms have to function at the same time in the interior of the brain. And when I say two, I am probably falling far short of the truth. We are evidently amazed by this complexity, and then forced to suppose that an enormous number of genes must control the formation of the eye. It is probable indeed that this number is considerable but we do not know how to guess the total.

Haldane has said that the number of genes in a mammalian cell can be estimated at 40,000, but you will note that this evaluation is unprecise—I will not say fanciful—for Haldane was a serious scholar. This is about all we can say, for we only know the genes which undergo mutations. We ignore the others.

We know about 1,500 of the genes in the human species. There may be many others that we do not know. In *Drosophila*, we arrive at ten or eleven thousand genes by extrapolation. In man we do not know; perhaps 100,000, perhaps many more, perhaps many less. There may be combinations of genes which allow a determinism and control certain characteristics.

We then run up against an extremely difficult question. How shall we explain how these variables organized themselves? I do not use the term "mutation" so as to avoid dealing with any theory. I do not wish to do so. There had to be a system of coordinated variables. It is, therefore, the problem of coordination which is in question.

If we admit that there is only one variable, we do not understand the question. But statisticians show us that it is extremely difficult for two variables which are produced at the same time to function. The calculation has been made: the probability of functioning is infinitesimal. It is perhaps of the order of 10^{-10} or 10^{-15}, therefore of the astronomical order.

III.

When one reaches such feeble probabilities, I think one must abandon them and look elsewhere. Let us consider the human brain. The human brain is not at all, as had been thought, the result of a mutation which would have doubled the number of neurons. Take the brain of a gorilla and multiply it by two and you will have a bigger gorilla brain, but it will always be the brain of a gorilla. Moreover, it is not the biggest primates, the *Gigantopithecus* and others, which have survived. Not at all. Bigness is not a factor for success, there is something else. There is a modification not only of architecture, but there is also an acquisition of new characteristics.

We are beginning to know a great deal about the science of heredity. Genetics has made enormous strides and molecular biology has consolidated this progress and contributed greatly to defining our knowledge. We can say that we know *grosso modo* how characters are transmitted and we can even say, from an immediate point of view, why they are transmitted under such and such conditions.

What we know is that heredity, namely stability, is the dominant rule, but that this rule is subject to variations. There are in fact accidents in the constitution of genes; but we will not enter into the system of the genetic code, of its polyploids, or others. It would be beyond the scope of our considerations.

The great problem for an evolutionist is to know how these innovations appear. We know how a gene can be modified by mutation. We know how it is transmitted from father to son, from mother to son, but we have no idea how the gene itself is acquired.

What we know, and it is not encouraging, is that bacteria of the type of the colon bacillus, the famous *Escherichia coli,* the battle horse of all molecular biology, is that these bacteria contain in one unit 300 times less deoxyribonucleic acid (DNA) than the cell of a mammal. This is an enormous proportion for a bacteria which is evidently complex and disposes of an important chemical arsenal with all its enzymes. But what is it beside a man? And it seems that our cells do not contain much more than 300 times more DNA. There is something here which we find difficult to understand. Things become even more difficult when we take into account that the protozoa or protophytes or radiolaria have enormous quantities of DNA.

I know that one can get around the difficulty. We have a global quantity of DNA for a given cell, but this does not mean that the same combinations of DNA are not repeated. In other words, the same chromosome A can be represented two or three times in the cell with the same sequence and the same chemical consitutents.

But there are cases when this rule cannot be invoked, because the description of the chromosomes is not very well known. What one calls the "karyogram" and the chromosomes do not repeat themselves. Therefore this argument is not wholly valid. It is not refuted, but a great deal of good will is needed to accept it.

We do not know how genes are acquired. We can suppose that the primordial flagellate, the primordial amoeba, contained all the genes and it lost them, or else that the genes were all in a potential state and that they could only develop their potentiality in proportion as the cytoplasm was modified.

But this hypothesis is above all incompatible with the figures and with the ponderable evaluations that I just pointed out. Furthermore, it is absurd; it is giving away to a frightful alienism to pretend that this interlocking of germs and of genes could explain what we are. It is really a question of acquisitions.

Moreover, evolution is essentially acquisition. It is always that which is new. It is the passage of one plane of organization to another that we must explain. On this plane—I am categorical—we are ignorant of everything. We do not know how we acquire new genes, nor how a gene establishes itself or inserts itself into the chain of the existing DNA. We are ignorant of all of this.

IV.

I will now take you into another realm. Many animals on earth have never evolved since very ancient times. First, we have the bacteria. Bacteria are known to have existed for about 1,200 million years. It is possible to date the first bacteria, and the blue algae are of the same age. One of the filamentous blue algae that are built like bacteria, the "stromatolites," reproduce themselves by simple fission. They have not changed. There is also a kind of *Oscillatoria* (they are not exactly *Oscillatoria*, but algae that resemble them closely) dating from the ante-Cambrian periods and which can almost be classified in extant genera. Actually there are also brachiopods that have not changed either. We even have mammals that have not changed: the daman, for example, has not changed; and for the most part, the insects have not budged.

There seem to be new species among the insects but if we take a given group, the termites, we ascertain that they have existed a very long time. The termites found in Baltic amber, in the Oligocene, namely 60 million years ago, are the same as those of the present. The genera are the same, they can be determined and recognized. The most evolved forms are present. Therefore there have been only small variations since that period.

But let us consider the kitchen cockroach, which is particularly interesting. One always talks of living fossils, but the most beautiful living fossils anyone can talk to you about in Paris are the kitchen cockroaches. They have remained practically unchanged since the end of the Carboniferous, since the Permian. Its evolution is minimal, it would simply affect the copulating organs; but the wings have not changed, and the general anatomy is identical.

And yet–I stated long ago but the facts have only just been confirmed–these very ancient beasts undergo *mutations* like the most recent forms. But they have not *changed*. This point must be underlined. How is it that cockroaches which have a very high rate of mutations do not become different? There are no modifications.

The horse has arrived at the end of evolution since it is a recent animal. Yet it presents a sizable number of mutations, from the Shetland pony to the native of the Ardennes, or from the Boucheron to the Percheron. The horse, however, does not change. We have slight changes but they are embellishments. The plan is unchanged. I would certainly not say that evolution has ended. Yet should you ask me to demonstrate that evolution continues, I would be hard put to do so.

One can perhaps speak of certain kinds of fish. My friend Theodore Monod could enlighten us on this point. Or one might speak of certain kinds of rodents, but this does not take us very far.

You might object that our life is much too short for us to be able to perceive the changes, but it seems to me that we should be able to ascertain at least a few. Our paleontological documentation is very rich, yet we see no (or at least very few) changes. Certain authors, including my predecessor Maurice Caullery, have written (not without a certain courage) that they think evolution has ended. I do not think so. Evolution still goes on, in my opinion. Is it perhaps a residual evolution?

If we now return to the position I indicated at the beginning, if we take a very high position—that of a satellite—we ascertain that with respect to the evolutionary fresco, no new plan of organization has appeared since the Triassic, the beginning of the secondary period. The mammals appear then in regular succession to the reptiles with all the conditions of passage perfectly prepared. But nothing has appeared since.

V.

And yet there is a great novelty and that novelty dates comparatively speaking from yesterday: our brain. The brain has modified evolution; it has introduced something entirely new, something which has, so to speak, changed living phenomena on the surface of the globe.

Man has only lately begun to understand what his brain represents in organic evolution. It is a phenomenon as important as the formation of the pluricellular state. Without a pluricellular state evolution was in a blind alley! The unicellular being cannot go beyond a certain evolutive stage. It can make certain experiments, certain attempts. When, for example, you study the ciliary infusoria or the radiolaria you have an infinity of forms, but they are always repetitions, embellishments. The plan does not change and new functions do not appear. A determined number of cells was needed so that the organism could evolve. To become complicated it had to have not two but three germ layers; this third germ layer being preeminently the organogenic germ layer.

You see then that in evolution there are phenomena which are indisputable and which raise problems when the evolutionist confronts them. For my part I accept a great deal of neo-Darwinism. But what annoys and irritates me is to hear some neo-Darwinians say that everything has been explained; we can now lie down and rest.

On the contrary, everything is only beginning to be explained. The neo-Darwinians do not understand that we are at a stage in biology when the most important thing is to describe with exactitude the phenomena and to at least look for their parameters, their causes. As long as this analysis is not made we cannot pretend to explain anything. We can only measure with badly defined units phenomena which in reality we have no possible means of measuring. It is on this plane that I find that neo-Darwinism goes too far, though in other respects I have a great deal of esteem and even admiration for it.

VI.

But as I told you a few moments ago, there are sensible neo-Darwinians, and if you will permit me I would like to add a few words to what Chauvin has already said, and show you how the problem of finality appears in biology at certain moments.

There are cases when all living phenomena show first of all a subjacent finality—a finality which is perhaps not a transcendent finality but which is in a way an immediate finality. But I will try to adduce certain difficulties of the problem of finality in biology.

A bird still in the egg is in an extremely difficult position. The egg shell poses an obstacle, but although a certain phenomenon is produced—namely, that the thickness of the shell is decreased by a chemical process—emergence from the egg still is a problem.

It turns out that the chick has everything it needs in order to emerge. Oh, he is feeble, this chick, even when it is a question of birds whose chicks are what we call "nidicoles," poor little things, naked creatures, whose muscles are flabby and badly developed. Even this chick has all the necessary tools to liberate himself. He has the proper muscle and innervation. It is not at all an ordinary muscle; it is a muscle which inserts itself on the nape of the neck, on the occipital and on the first vetebra. Then, thanks to this muscle, the birdlet strikes light hammer blows until the shell breaks; he strikes with the very hard point of his "diamond."

But there are also cases where the shell is very thick. Take the case of cuttlefish. Its shell is of a horny nature like rubber. There the problem is very difficult. There is no possibility of a mechanical opening. It is impossible. But the young cuttlefish possesses on its dorsal side a gland whose enzymes are perfectly adapted to digest the shell.

Imagine the problem bristling with difficulties that a mulberry butterfly has to face when trying to hatch! And you will be still more edified when you know what a cocoon is. The cocoon is not only made of multiple vests of silk thread, but the most internal ones form an extremely resistant coating, called the telette, and which is composed of more or less sclerous, structural proteins. The butterfly has no mandibles. It has no hatching "tooth." It should not be able to emerge. However, at the end of pupation the young adult starts to manufacture a saliva which is extremely alkaline. He lifts himself up on his legs and spits on the superior pole of his cocoon. Under the influence of the alkali the strands are digested. But this is not all. With his forelegs he separates the strands magnificently and he is out. This is how the problems of finality are put to us.

It is possible that the neo-Darwinians are right and that it is a pseudoteleology. It is possible. But these neo-Darwinians who say they have torn finalism from their concepts forget one thing: namely, that finality is the couch grass of biology. The more you pull it out the stronger it grows. And do you know who gave strength to finality? The neo-Darwinians themselves.

What have they brought us? They have brought a living universe in which there is a motor, namely, utility. Nothing that is useless persists. Only what is useful develops, prospers, and perpetuates itself. If utility does not have a finalist subfoundation then I surrender.

The Darwinian universe is the most finalized universe there is. Even a strict Lamarckian would never have spoken in those terms. The book by Dobzhansky (for which I am to write a preface) contains, precisely, a chapter entitled "Of the Principle of Utility." I will say no more!

Man's Place in Nature

Morris Goldman

The title of this talk is taken from Thomas Henry Huxley's book of that name, which appeared in 1863. To judge from the record, the subject appears to possess a permament attraction for essayists and lecturers.

The various opinions concerning man's place in nature can be grouped into two broad categories which I shall call the religious and the secular. Leaving aside questions of detail about which there may be considerable differences of opinion, the religious outlook is characterized by the view of man as a transcendental creature who has duties and privileges that go beyond what is applicable to the rest of nature. Furthermore, and perhaps more basic, is the concept that there exists somewhere, somehow, a God-head which in some way controls and directs the natural world which we see, although it is itself beyond the control of natural laws.

The secular view denies both these propositions. Stated more positively, it sees man as one animal species among millions of others, with no inherently special privileges beyond what it makes for itself by virtue of its unique mental capabilities. There is no supernatural God-head, and whatever happens does so in accordance with universally applicable natural laws, all of which are potentially, at least, understandable by man.

The religious viewpoint, as given here, is obviously based on certain undemonstrated (in the scientific sense) and unprovable axioms. Secularists like to believe that their world outlook is coldly rational and built on a solid foundation of demonstrated truths. I shall try later in this talk to show that that is hardly the case, and that secularism demands much faith to make it viable intellectually, just as does the religious outlook.

Efforts have been made by outstanding scientists like Lecomte du Noüy, Teilhard de Chardin, and E. W. Sinnott, to construct secular, scientific frameworks upon which to build a religious view of man's nature. With all due respect to the profound thought and effort that has gone into these attempts, I am myself convinced that no true synthesis is possible at the present time. That is to say, there is no evidence or logic available to either side which is so overwhelmingly convincing that all serious opposition must inevitably submit, or so innocuously blend as to be absorbed by the other side with minimal disruption of its own principles. For example, speaking now specifically of the traditional Jewish view of the world, which I am sure has parallels in other religious philosophies, divine guidance of the world is an absolutely fundamental theorem to the faithful. When this is coupled to another article of faith, that the divine Creator has no shape or form and is free from all properties of matter, then it becomes obvious that divine intervention in nature and in human affairs is never demonstrable in scientific terms. Such views are utterly beyond accommodation in a secularist frame of thought.

The secularist view that man who worries about right and wrong, who writes music and poems, who plots the stars and plumbs the oceans, who feeds the needy and slaughters his brother because of differences in philosophical abstractions, who decorates both his hovels and his palaces so that they will be more pleasant to look at, who buries his dead and wonders where they have gone (and so on and on through the list of uniquely human characteristics), the view that this man is at one with, say, the tapeworm, in being no more than a product of blind, evolutionary forces, is to the religious camp so patently a black misreading of reality that it can only be rejected as absurd.

The main purpose of my talk, then, is not to attempt a reconciliation between these diverse views, which I consider impossible, nor to present arguments favorable to my personal religious outlook. Instead, I would like to consider aspects of one of the dilemmas faced by secular humanists, with the intent of demonstrating the difficulties to be found in the bases of this naturalistic philosophy. Of necessity, the argument will be brief and incomplete because of time limitations.

The dilemma of the secular humanist is this: if all of nature is planless as to its ultimate end and, therefore, purposeless, and if man's unique mental and spiritual equipment is just another example of blind evolutionary diversification, then man, like other animals, has no planned entry into this world and no planned function to perform now that he has arrived. In that case, how does the secular humanist justify his well-known concern for man's ethics and morals, both concepts implying goals and standards, and thus plan and purpose? Put another way, is the question "Am I my brother's keeper?" a valid question, naturalistically speaking? If it is, does there exist a humanistic answer that can be deciphered from an objective examination of nature, without any assists from revelatory religion?

The full difficulty of the problem becomes apparent only after dismissing from our minds the Western ethical system in which we have grown up, with its notions of right and wrong, and good and bad. This is necessary because the Jewish and Christian thinking which forms the core of Western ethics, is based on revelation, and revelation must be completely rejected as a source of knowledge in the secularist framework. To discover a truly naturalistic ethic, if one exists, we must look at the world through the eyes of an amoral viewer, unencumbered with preconceived standards of behavior.

Bentley Glass, in an article titled "The Ethical Basis of Science" formulates a very biological answer to the question. He says:

> The evolutionist is quite prepared to admit the existence of right and wrong in terms of the simple functions of biological structures and processes. The eye is for seeing.... Sight conveys information about food, water, danger, companionship... and other vitally important matters. Should one not then say "To see is right, not to see is wrong"?[1]

In this remarkable statement, Glass, in effect, makes the assumption that (in naturalistic terms) it is better to survive by using one's senses, than not to survive—say by ignoring one's senses. But in the history of the world, death and extinction are at least as prominent as life and evolution. In the economy of nature there is no indication that it is preferable for an atom of iron, for example, to be part of a hemoglobin molecule in a living animal, than to be part of a ferric oxide molecule in some mineral deposit. Thus Glass' first assumption is not at all implicit in the facts of nature, but is rather an intuitive, human determination that life is better than nonlife, and therefore what promotes life is better than what promotes death. In all objectivity, this assumption cannot be made even if we start our search for an ethic with the world of life alone. Curators of zoological gardens know that there are many animals so constituted that they will starve to death rather than accept perfectly nutritious food which, however, is different from what they are accustomed to in their natural condition. And if such animals die, is there any sign of mourning in nature that these creatures have returned their substance to the eternal cycle of life and nonlife?

But even granting Glass his assumption as naturalistic, which it is not, extrapolation of his principle leads immediately to what are —ethically—somewhat horrendous results. For example, if the eye is for seeing and seeing is good, then, since the claw is for killing, killing must also be good. In a hunting animal with a well-developed brain, like man, weapons are extensions of tooth and claw; therefore mental activity to improve weapons for killing is natural and good, too. It follows, then, that the Germans, for example, exercising their biological right to improve their conditions as they saw them can hardly be condemned on naturalistic grounds for murdering six million Jews and four million assorted other nationals, since they were simply using their natural gifts in this direction to the fullest extent. The clinching evidence that the German behavior must have been "right" in Glass's naturalistic terms, is that Germany emerged from the bloodbath with a higher potential for survival than any of her victim peoples.

Actually, of course, Glass does not extrapolate in these terms, choosing instead to call "good" only those aspects of mental activity which we would all applaud as good and right. But the basis for his choice cannot be read out from an examination of biology. Rather, he is simply reflecting his heritage of revelatory Western ethics.

Simpson rejects all efforts to derive ethical systems from an examination of nonhuman biology since, in his view, evolution is a completely amoral process. However, since man, by some evolutionary quirk, does possess a moral sense, Simpson feels constrained to create, rather than derive, an ethical syllabus based upon his own personal evaluation of what is important and unique about man's nature. Thus, to Simpson, acquisition and promotion of knowledge, personal responsibility, and heightening the integrity and dignity of the individual, form the cornerstones of an ethical edifice suitable for *Homo Sapiens*, as he now exists.

The fatal flaw in Simpson's program is his determined effort to keep one foot planted firmly in each of two irreconcilable positions. On the one hand he states: "These ethical standards are relative, not absolute. . . . They are based on man's place in nature, his evolution, and the evolution of life, but they do not arise automatically from these facts. . . . " Furthermore, he states: "He [man] can choose to develop his capacities as the highest animal and to try to rise still farther, or he can choose otherwise. The choice is his responsibility, and his alone." On the other hand, Simpson also says, "Authoritarianism is wrong. . . . This is an ethically wrong denial of the personal responsibility inherent in man's nature. . . . Totalitarianism is wrong. The concept of a state as a separate entity with its own rights and responsibilities contravenes the biological and social fact that all rights and responsibilities are vested by nature in the individuals that compose the state."[2]

Thus Simpson would propose an ethical system based, presumably, on man's natural history, and at the same time disclaim any authority for his system in the case of anyone who chooses to see man's nature in different terms. As an ironic aside, at this point, we take note of the fact that the one nation in the world which is most overtly secular and Darwinian in its view of life, the Soviet Union, is also among the most completely authoritarian in its politics, and is one in which acquisition and dissemination of truth hardly holds an honored position in its activities.

From Simpson's view that nature is amoral, and that man evolved accidentally, it must follow that he cannot be charged with obligations to the rest of nature or even to his own species. Thus, any behavior that will somehow satisfy his accidentally evolved sense of right and wrong should be acceptable, speaking naturalistically. Therefore, the artist who views creativeness in man as more important than acquisition of knowledge, the demagogue who recognizes the attraction to the masses of strong and disciplined leadership, the Communist who places the welfare of the community above the welfare of the individual—all are equally as entitled as the distinguished professor to create systems of behavior that would maximize those uniquely human attributes which they favor. Simpson's own ethical proposals are wonderfully American early twentieth century vintage; but by his own estimation, they are not implicit in man's nature. It seems obvious that they reflect Simpson's American Christian heritage much more than any naturalistic, evolutionary determinism.

The least naturalistic and most metaphysical attempt to provide a nonrevelatory ethic for man is made by Julian Huxley in a rather long essay, "The Humanist Frame." In this often lyrical, sometimes sermonlike paean of secular humanism, marred here and there by poor biology and fuzzy politics, Huxley avoids the dilemma we are considering by simply ignoring one of its horns. Thus, he takes for granted, without any reasoned justification, that although man is part of a comprehensive, unplanned evolutionary process, he cannot avoid playing a decisive role in this process. Presumably, this is due to man's superior intellect, but Huxley offers no justification for his decision to place this intellect at the disposal of nature as a whole, rather than restricting it solely for man's own aggrandizement or for some other purpose for that matter. Huxley says: "Man's destiny is to be the sole agent for the future evolution of this planet."[3] A tall order for a race of saints, not to speak of a congregation of secular humanists.

In this respect, then, Huxley departs widely from Simpson's point of view, offering man not a choice but a burden. Having made this personal decision, Huxley proceeds to outline a rather predictable idea-system which may be briefly characterized as Christianity without Christ. That is, he calls for the constant improvement of man as an individual and in society, for permitting man full artistic and creative fulfillment, for surrounding man with beauty and love, and so on, including even the ceremonials of religion—carefully emptied, however, of any notions of God.

It appears then, from this brief examination of the three approaches to naturalistic ethics, that nonhuman nature does not present us with an ethic that would be acceptable to most of us, and that humanist proposals from Western professors of biology have more in common with Jewish and Christian revelation than with ineluctable conclusions implicit in a Godless universe. In fact, one often has the impression of being offered revelation without religion by these prophets of secularism.

We seem thus to be left with our original dilemma: morals for man in an amoral universe. The heart of the difficulty, of course, is the secularist notion that man really is only another unplanned life-form, endowed only accidentally with his unique social and intellectual qualities. This notion is so contrary to the practically universal, intuitive feeling that man represents something more than purposeless existence, that the evidence offered in its support deserves the most careful scientific scrutiny before it, and the problems it brings in its wake, can be accepted. Let us therefore examine some of the evidence.

According to widely accepted views current at the present time, man's evolutionary history within the mammals is said to begin with prosimian ancestors living in the Paleocene-Eocene epochs, some 60–70 million years ago. These smallish animals, similar in appearance to modern lemurs, lorises, and tarsiers, have left fossil remains in Europe and North America. For reasons not really understood, these animals disappeared from the fossil record before the close of the Eocene, reappearing again in the Pleistocene, about 50 million years later. The morphology of modern representatives of these groups appears to differ little, if any, from their presumed ancient precursors.

Between the prosimian primates of the Eocene and the apelike fossils from the Miocene, about 15 million years later, only a small handful of teeth and fragments of jaw have been found, all in a single locality in northern Africa, in what is now Egypt. The relationships of these few fragments to any other animals need to be inferred without benefit of any supporting information concerning cranial volume, limb structures, or living habits; essentially all conclusions are based on dentition alone.

About 20 million years ago, in the Miocene, there lived in Europe, Africa and Southern Asia species of primates which left fossil records, mostly in the form of assorted tooth collections. Pieces of jaw and limb bones also occur, sufficient to assign the remains to ape, rather than monkey sources. Except for the skull and some fragmentary bones found in East Africa, to which the name *Proconsul* has been applied, these dental shards are essentially all that are available from this period. Romer states concerning the Miocene fossils: "Our knowledge of the fossil history of these higher apes (i.e. gorilla and chimpanzee) and of presumed human ancestors on this level is tantalizingly poor. . . . "[4] These fossils have all been placed in the genus *Dryopithecus*, and this ghostly assemblage of teeth has been set up as belonging to the probable ancestors of man.

Continuing with this Hominid lineage, i.e. the group that includes the modern apes and man, we encounter another gap in the record of about 20 million years between the Miocene genus *Dryopithecus* and the more recent fossils of the Pleistocene. This gap is relieved only by a few teeth and jaw fragments from India and Africa, assigned to the genus *Ramapithecus*. Nothing is known about the cranium or limbs of the creature that left these teeth, but on the basis of dentition alone, *Ramapithecus* has been placed in or close to man's ancestral tree.

The fossil record of the Pleistocene epoch, which includes the last one to two million years, reveals in relatively rapid succession the Australopithecines, *Pithecanthropus*, Neanderthal man and, starting about 30,000 years ago, modern man. *Australopithecus*, found in South Africa, possessed massive projecting jaws equipped with great but humanlike teeth, and appears to have been an erect, bipedal anthropoid. Nevertheless, the volume of his cranium was only about 550 cubic centimeters, which is within the range of the chimpanzee and gorilla, and less than half that of man.

Pithecanthropus, called by some *Homo erectus*, has left remains primarily in eastern Asia, comprising so-called Peking and Java man. Dentition and stature of this group were manlike but, as in *Australopithecus*, the face was projecting and chinless in the apelike manner. The braincase was low with very heavy browridges but was a bit less than twice the size of the South African primate. There are indications that *Homo erectus* used simple stone tools and fire.

During the last glacial period of the present epoch, *Homo neanderthalalis* made his appearance, remains having been recovered mainly from Europe but also from the Middle East, Asia, and Africa. Although Neanderthal man had a brain as big as or bigger than that of modern man, the appearance of his skull was rather different; the supraorbital ridges were very heavy, the forehead was low, and the large brain size was achieved by enlargement of the back part of the skull; the chin was formed rather weakly.

Relics of a primate who, in all ascertainable physical respects was comparable to modern man, are found starting about 30,000 years ago. The forehead is high, the nose and chin strongly formed, and the teeth and jaws of manlike size and appearance. Olson remarks that the rather dramatic and sudden appearance of this species, which bears our own name, has left some doubts as to where *Homo sapiens* came from. It would seem to him that *Homo sapiens* developed somewhere beyond the range of present finds of fossil men and then, fully matured, penetrated rapidly into the lands of the Neanderthals, replacing this less advanced type during a relatively short span of time.

By 15 or 20 thousand years ago this man was painting the walls of his shelters, perhaps in connection with cultist practices related to hunting, and his cultural progress can be followed in broad terms up to historic times. This man, therefore, who appears suddenly and discontinuously as a full human, is the direct ancestor of present human populations.

It appears to me that an objective, nonprejudiced reading of the fossil record of the primates hardly confirms the story, repeated endlessly in popularized accounts, of clearly defined transitions from ancient lemuroids to tarsioids to apes, from apes to man-apes, and from the latter to modern man. On the contrary, the story reconstructed from the tangible evidence, as opposed to speculation, is one of tremendous gaps in the record both of time and space, of tenuous grasping at teeth to establish phylogeny almost to the exclusion of the rest of the body, and of gross discontinuities between forms supposedly related to each other by direct genetic descent. Responsible paleontologists readily admit in technical print the difficulties of interpreting the actual record. Nevertheless, secularist interpretations are offered freely, and we need now to look at the reliability and precision of the interpretative methods which are used.

A first point which we cannot avoid raising at this juncture concerns the entire concept of the mechanisms of evolutionary diversification, as presently understood. To quote Mayr on current theory:

"The proponents of the synthetic theory maintain that all evolution is due to the accumulation of small genetic changes, guided by natural selection and that transpecific evolution is nothing but an extrapolation and magnification of the events that take place within populations and species."[5] In terms of the magnitudes of the effects being dealt with, it may be proper to compare this statement with one claiming that the explosion of a nuclear device, for example, is entirely explainable by simple extrapolation from what is known about the detonation of sticks of dynamite.

Strain differences within species commonly involve changes of single base pairs in the sequences comprising the genetic DNA of the species. Man, as well as many other vertebrates, possesses on the order of 5 billion nucleotide pairs per diploid cell. If the matching of sequences between man and other primates averages about 90 percent, as appears to be the case from the *in vitro* hybridization work of McCarthy, Hoyer, and their associates, this means that 10 percent of the sequences are different; or in other words, differences in 500 million nucleotide pairs have had to accumulate in the DNA sequences of the species involved since man and the other primates diverged from a hypothetical common ancestor. This figure is perhaps subject to change by a factor of as much as ten, depending upon technical details which as yet remain to be worked out, but it gives one an idea of the range with which we are dealing.

To presume, without any direct evidence at all, that differences of such magnitude accumulate in a simple additive manner, in the same way that differences in a few nucleotides accumulate between isolated strains of the same species is truly to assume a most unscientific posture. A fact of significance in this regard is that no examples in nature or in the laboratory of the evolution of one species to another have ever been observed or induced. The single exception occurs in instances of polyploidy in plants where the chromosome number is doubled or otherwise altered as the result of a defect in the production of germ cells. These cases are rare and have never been considered a significant factor in evolution in general. One would not expect in our limited lifetime to see terrestrial carnivores transformed into whales, but creation of new species or higher taxa among the bacteria or protozoans should be within the reach of the experimental laboratory, if the assumption quoted from Mayr holds true.

Dependence on tooth structure is so strong in accounts of primate evolution, that it is pertinent to ask how well primate tooth morphology reflects the morphology of the rest of the animal. Simons says: "It is of considerable interest that the skeletal material of *Pliopithecus* now available shows that although definably hylobatine dentally (i.e. similar to *Hylobates*, the modern gibbon), the forelimb elongation so characteristic of modern gibbons is barely noticeable in this Miocene form."[6] The metaphysical question of whether it is the teeth or the forelimbs that make a gibbon need not concern us; what is important is that one is not necessarily a guide to the other. In fact, the accepted evolutionary principle of mosaic evolution which holds that different organ systems may and often do evolve independently of one another, would caution us against drawing general conclusions on the basis of a single system alone. Nevertheless, practically all speculation concerning the ancestry of man up to the Pleistocene is based essentially on scattered tooth remains only.

In the absence of the living animal or other evidences, the habits of ancient species must be inferred from the structure of their fossilized skeletons. Observations of living primate species show that such inferences may be widely in error. For example, the gorilla, which is anatomically a brachiating creature, fitted for an arboreal life, is in real life a terrestrial species usually getting around in a modified quadrupedal manner. The gibbon, which is highly arboreal in habit, with forelimbs specialized for acrobatic swinging among the branches of trees, is also the most adept of the apes at walking in an erect, bipedal manner.[7] The implications for speculation concerning phylogenies and natural selective pressures of these confusing associations of one kind of anatomy with another kind of habit, hardly need any further emphasis.

With what precision and on what grounds can one place a particular animal in the evolutionary progression from primitive to advanced status? It is generally assumed, for example, that monkeys represent a more primitive primate condition than the great apes. But the fossil record of Old World primates shows the presumably more advanced apes appearing earlier in time than the presumably more primitive monkeys. Old World monkeys first show up in the record, and then only in limited numbers, in the Pliocene, about 10 million years *after* the appearance of the apes. Romer has the following to say about the New World family of little monkeys, the marmosets: "A peculiar feature in all except one marmoset is that the last molar has been lost, the only case of its complete reduction among primates (although man is approaching this condition). . . . It has been thought that the marmosets are the most primitive of monkeys, but features such as the molar loss suggest that they are specialized rather than primitive."[8]

A perennial problem in constructing phylogenies is to distinguish between common ancestry and diverse ancestry followed by parallel or convergent evolution. Simons comments concerning some extinct primate lines:

> The common qualities of tarsiines and necrolemurines could be attributed to parallel evolution as was implied by Hurzeler. However, if common characters of the level of frequency seen between these two groups be interpreted as parallelisms then it would probably be impossible ever to sort out the difference between parallelistic and common heritage characters, and the study of phylogenetic trees would wither at the root.[9]

This is a strong statement, indeed, in the light of the status of New and Old World monkeys.

Most visitors to a modern zoo, even if they are observant, would find it difficult to distinguish between Old and New World monkeys unless certain anatomical features were called to their attention. Physical similarities between the two groups far outweigh the differences. This remarkable resemblance exists despite different origins, presumably from different prosimian ancestors in Europe and North America, and despite complete separation by vast oceans for a period of about 60 million years. Not only do these two groups resemble each other morphologically, but they also show similar behavior patterns. Andrew says: "The parallel evolution of similar displays in the Cercopithecoidea (Old World monkeys) and the Ceboidea (New World monkeys) has already been remarked on. The resemblance of both to the displays of *Canis* (i.e. the dog family) are even more remarkable in view of the far more distant relationship of *Canis*."[10]

Thus, parallel evolution or common heritage is invoked not on the basis of objectively clear distinctions, but rather on the basis of what will best fit a previous speculation concerning the relationship between the forms involved. In the case of the ancient primate record, there is so little tangible evidence to go on, that the choice between the two alternatives becomes a very subjective one, indeed.

I do not think one needs to be a hard-shelled, religious fundamentalist to express skepticism concerning the account of man's descent from the apes, as the story is currently being told. The gaps are too great, the evidence too fragmentary, the interpretive methodology too freewheeling to feel smug about knowing even the main outlines of the story. But suppose the main outlines of the very ancient history of man, as told by the secularists, are conceded. Does this now place the more recent story on the solid, scientific foundation which the secularists like to claim? Unfortunately not, for then we must account for the phenomenal increase in the size of the brain in a most remarkably short period of time.

According to present speculation, a period of relative stagnation in brain size lasting at least some 20 million years, was followed by an explosive increase in cranial volume of two- to three-fold during the past 1 to 2 million years. Speculation abounds concerning the reasons for this sudden and unprecedented increase. Le Gros Clark writes: "The demand for skill and cunning in arboreal life was, no doubt, one of the reasons why the brain began to expand in size and complexity very early in the evolutionary history of the Primates."[11] One thinks, of course, of the South American sloth, almost exclusively arboreal, whose very name is synonomous with qualities opposite to those we associate with bright, agile primates. In addition, arboreal squirrels, known as fossils since Oligocene times (about 30 million years ago) do not show any conspicuously greater intelligence than the varieties inhabiting terrestrial niches. In any case, the tremendous cranial expansion of man supposedly took place in a bipedal, ground-inhabiting creature, removed from arboreal life by tens of millions of years.

Some attribute the increase in brain size to strongly selective feedback effects resulting from the social nature of monkey and ape life. Mayr thinks that the need for more efficient communication, like speech, accelerated the rate of brain development.[12] Yet prosimians like lemurs, with a fossil history of 60 million years, live in monkey-like societies involving long youth and social learning, but have not developed anything as high even as monkey-level intelligence.

Any argument concerning the selective advantages of increased intelligence must take into account the fact that these advantages should accrue to most or all animals, or at least mammals. It appears that modern placental mammals, on the whole, possess bigger brains than mammals of the Eocene epoch (Jolly, *Science* 153:504). But cerrtainly there is nothing in nature to compare with the kind of change which occurred between *Australopithecus* and *Homo sapiens* in the brief period involved.

There are other evidences of the fundamental ignorance that surrounds the origins of man's peculiar characteristics. For example, all other primates grow a more or less luxurious pelage which is often the subject of self- or fraternal-grooming. Man obviously does not conform to this pattern, and it is not known how or when the reduction of his body hair occurred. In an exchange of letters in *Science* during 1965–1966, Bentley Glass and five correspondents offered six different and mutually exclusive speculations concerning the origin of this unique trait in the order of Primates. Concerning the origin of speech in man, Marler writes (*Science*, Aug. 18, '67, p. 769): "There is still no plausible explanation for the emergence of the cultural transmission of patterns of sound production in man."

These questions I have raised concerning the validity of the evidence and interpretive methodology used by the secularists to construct their picture of man's place in nature certainly do not disprove the secularist story. On the other hand, in my opinion, they provide a sound and rational basis for rejecting as unproved, and perhaps unprovable, the fundamental secularist assertions that physical man emerged accidentally from some pool of monkeylike ancestors, and that his human qualities resulted from accidental favoritism by a never clearly defined process of so-called natural selection.

It seems to me that the dilemma of moral man in an amoral universe remains unresolved for the committed secularist. But for those who have not yet adopted the secularist faith, I believe there are adequate scientific grounds for refusing to accept fashionable speculation as fact, and spuriously scientific ethics as a guide for human behavior.

You will forgive me, I hope, for closing on the following homiletic note. It is a stylish tune among secularists that evolution teaches that all things must change, therefore we must reshape our beliefs in the contemporary way. This tune ignores a different song, played on different occasions, that bespeaks the permanence and universality of life mechanisms. For example, the genetic code, the ultrastructure of cilia, striated muscle, respiratory pigments, the complex of cellular enzymes involved in energy utilization, and in other cellular activities, these are just some of the fundamental configurations that are associated with life at all levels of organization. The permanence and ubiquity of such key substances and structures leads us to believe that they cannot be permitted to mutate in any substantial manner without a prematurely lethal outcome for the unfortunate mutant.

Would it not be ironic if the concept of transcendent religion which is so universal in all societies were to turn out to be one of the fundamental configurations that make truly humanistic society possible? And would it not be ironic if the secularist mutation should turn out to be the lethal factor which would strip man of his humanism and turn him truly into the animal which the secularists seem to desire him to be? Even the nonreligionist may well pause and ponder before proceeding to trade 4,000 years of rich religious humanism, dating back to the biblical Abraham, for the thin mechanistic gruel of the secular evolutionists.

NOTES

1. Bentley Glass, "The Ethical Basis of Science," *Science,* Dec. 3, 1965, p. 1254.
2. G. G. Simpson, *The Meaning of Evolution,* III (New Haven, Conn.: Yale University Press, 1949), p. 281.
3. T. H. Huxley, *Man's Place in Nature* (1863; reprinted, Ann Arbor, Mich.: University of Michigan Press, 1959).
4. A. S. Romer, *Vertebrate Paleontology,* 3rd ed. (Chicago: University of Chicago Press, 1966), p. 224.
5. E. Mayr, *Animal Species and Evolution* (Cambridge, Mass.: Harvard University Press, 1963), p. 586.
6. E. L. Simons, "A Critical Reappraisal of Tertiary Primates," *Evolutionary and Genetic Biology of Primates,* ed. J. Bueltner-Janusch (New York: Academic Press, 1963), p. 110.
7. E. P. Walker, *Mammals of the World,* 3 vol. (Baltimore: The Johns Hopkins Press, 1964).
8. Romer, *op. cit.,* p. 220.
9. Simons, *op. cit.,* p. 97.
10. R. J. Andrews, "The Displays of the Primates," *Evolutionary and Genetic Biology of Primates,* p. 302.
11. W. E. Le Gros Clark, *The Antecedents of Man* (Chicago: Quadrangle Books, 1960), p. 30.
12. Mayr, *op. cit.,* p. 635.

Finalism in Biology

Rémy Chauvin

Any discussion of Lecomte du Noüy manifestly cannot neglect the important problem of finalism which preoccupied him as a scientist and philosopher for so long a time. In consideration of the progress of the sciences and technology, however, and in conformity with the wishes of Madame Lecomte du Noüy, it is less important to make a profound analysis of the works of the man we are commemorating than to try to analyze the problem of finalism from the philosophical and biological point of view. I can do no more than cover very rapidly an extremely difficult and bushy ground.

First, I think we must attempt to give some preliminary definitions. They will be very vague and no doubt inexact; but when all has been said and done, they will be no worse than others.

It is out of the question to retrace the history of the notions of finalism, of finality or final causes. This would require going back to Aristotle—perhaps even further—and clearly we have not this kind of time. Let us content ourselves with saying that in living beings and particularly in man, it seems that we can ascertain the existence of an activity directed towards a constant goal—this is the important point—through the multiplicity of perturbations and situations encountered. It seems as if there were an aspect of finalized activity which starting from man spreads to all living beings and, therefore, to the whole of evolution.

Indeed, I think that the idea of finalized activity comes first from man, namely through introspection, just as the concept of cause, according to many philosophers, is born of the feeling of effort. It is, moreover, not evident that this is the only process which could have given birth to the concept of cause, for with the Semites it is no doubt the idea of generation which produced it.

But man is part of the universe and if there is finality in man, there must be some in the universe. This is an inevitable consequence which was well perceived by the strict, bigoted mechanists and which they wanted to exorcise at any price. It was essential to prove that on the human plane, finality (and therefore liberty, this is almost a necessary corollary) were only illusions even though their obviousness seemed to impose itself rigorously on artless common sense.

I am naturally extremely embarrassed by having to treat such a complex question in so short a time. I can do no more than raise the question of liberty. But it is nevertheless difficult to avoid it completely. I will, therefore, content myself with a consideration which seems to be a pirouette, but here I take refuge behind the authority of Whitehead who thought, as I also think, that a discussion aiming to prove that there is no finality would be senseless and that even to discuss the question of liberty by proving that it does not exist would in a certain sense be devoid of intelligible meaning.

The same argument applies to finality, to liberty, and to will. But let us leave the philosophical realm, where I am not very sure of my ground, and come back to the biological realm. I will cite the old saying—one of the few I remember from the Latin grammar of my youth—*Caesar pontum fecit*. Caesar built (or ordered built) a bridge.

Here the will of Caesar conditions the existence of the bridge, even though a hypothetical spectator from Mars observing the facts from his distant planet could describe the process of the building of the bridge in strictly mechanical terms, while considering the men as ants. It would not be inexact; it would simply be incomplete.

Indeed, common sense, crude as it is, shows us right away that if we suppress what our introspection teaches us to call the "will of Caesar," there can be no bridge, a thing which the man from Mars cannot know even though he can describe the process in terms of "incitements and inhibitions."

In a sense, we can say that we should avoid two inaccuracies: the pure finalist description, which would involve saying that the will of Caesar was in a way all powerful or, if you prefer, that the important thing to describe is the will of Caesar, the man Caesar. This description is no doubt inexact because it neglects all the complexities of the construction of the bridge. Caesar did not *build* a bridge; he *ordered* a bridge to be built and the pure finalist description neglects the "order."

The strict mechanistic description, on the contrary, starts from the absolute exterior (if I may use this expression, familiar to Ruyer) and this external description is also inexact. Or rather, it is incomplete. This strictly mechanical description may even constitute a subtle means of understanding nothing of the phenomenon.

We shall leave aside the question of knowing whether Caesar himself can be considered the product of a blind concatenation of materialistic causes. That is another question. Let us take care to interpret him well, this man Caesar, who has emerged from the cosmos. Let us take him as he is, without mutilating him and without mutilating his action in the name of philosophical a prioris. No matter what the genesis of Caesar, no matter what the mechanisms, moreover unknown, of the will, even if we accept the hypothesis of nonliberty we would observe it at the moment of the appearance of the phenomenon "evolution" which is followed or accompanied by certain external effects which can be described as "finalized activities."

This is always characterized by what philosophers designate by the barbaric term of "perseveration," namely the persistence of a certain direction through various means or accidents. This is important to underline because there are a quantity of rather new experiments on the behavior of animals which show conclusively that what often counts in the performance of an animal is not the act it performs in response to a stimulus but the goal it sets for itself.

From among a thousand possible examples I will cite a very curious experiment of this character. It concerns the rat who walks in a labyrinth and finds a reward in a box. If we take him by the skin of his back before he has gone through the labyrinth, put him in the box and make him ascertain that it is empty, and then place him at the entrance of the labyrinth, we perceive that his performance immediately changes. He makes all kinds of mistakes; the labyrinth no longer interests him.

This enables us to affirm that the important point in this experiment is the "expectation" of the rat. These English words are often strange to us but they correspond very well with what I wish to say. The important thing is not—as the strict mechanist thought it to be in the past—a more or less subtle combination (more or less well arranged on paper, of course) of stimuli and reactions. It is something that corresponds to notions which must be borrowed from the psychology of Dolmann, which can be called "mentalist." It is something very subtle which is expressed in para-human terms, that is to say "an expectation."

Now there are quantities of experiments which clearly show that in human terms a goal is actually pursued by the animal and that the methods it employs to obtain it matter little. The animal does not learn such or such a method in a preferential manner. He learns principally the general technique required to attain the goal.

Moreover, in the name of what would we decide to consider only half of the phenomenon in a human act; the half which is strictly finalized, namely human, or the strictly mechanistic half by decreeing that the other half is of no importance?

I think that if we do not wish to decide a priori we cannot escape the conclusion that the two halves are equally important, even though we have no very precise scale to show which one of the two gets the better of the other one at a given time. But it is unscientific, under the pretext that we do not understand the phenomenon "will" (I understand it no more than do others) to decree that the phenomenon "will" has no importance. There is here a very important difference.

On the other hand, a nonfinalist activity, a haphazard stroll in the woods, ends in human terms only in problematical uncoordinated results. Here I appeal to plain common sense which is perhaps the last word of science. When a man strolls by chance he does not walk in a preferential manner on one side or the other; he does not end by manufacturing a mechanism, a watch, for example. Therefore—and here I will present an extremely grave transition as bluntly as possible—the body of man and that of an animal are equally mechanisms, one would have to shut one's eyes not to see it. Is this mechanism, then, also finalized? Is it the achievement of an external worker who can only be God?

This is the question which makes the mechanists furious—and justly so, for it is presented here in outrageously simplified terms. It should naturally be expressed less crudely. At the end of this talk we will see how, in my opinion, it should be stated. But it must first be divided into several parts.

I believe that we cannot contest the fact that the organism in general is a mechanism, a machine, and not a series of problematical processes, all functioning separately. It would be a waste of time to try to prove the contrary. Indeed, it is sufficient to open a book on physiology to see that physiologists naively and without malice employ the vocabulary of finalism. Why? Perhaps, only because they cannot do otherwise and because to talk otherwise would end by making their accounts unbearably dull. Moreover, they do not ask themselves many philosophical questions on the subject of the language they use. But it is the language of engineers and what does an engineer do? He occupies himself with machines.

Once more I wish to repeat. The fact that opinions differ as to the genesis of the mechanism, by chance or by creation, is another question. But the fact remains that the mechanism exists; it must be interpreted. It must be looked at with a correct intellecual attitude. Contradictors will say: "Why do you need a finality for an organism? It adds nothing to the process and in this case you cannot look at it from the inside, as in man. You only see the organism from the exterior. Therefore you will never know what goes on in the interior of an animal. Your idea of finalism or of a finalized activity is neither an acceptable nor an applicable idea. It must, therefore, be rejected."

This objection is important and serious. I believe it is even the crux of the debate. Indeed, from the blunt point of view of the scientist, who perhaps does not sufficiently consider the shadings, what is the role of a scientific theory? It is very simple: the role consists in allowing a scientist to organize new experiments and to predict the results. A theory is a tool and nothing else; and when the tool is worn out, it is cast aside. Illustrious examples exist throughout the whole history of science. A scientific theory is interesting because it is useful, because it enables one to foresee the real. I do not use the word "true" because this is a word difficult to define. A theory must present a certain degree of conformity to the language which is measured by its applicability on the experimental plane.

Many scientists will say the idea of finality is not useful if it is not applicable. What more does it teach you than you would suppose without finality? But is this really so correct?

First, there are all kinds of ways of conceiving finality. We will come back to this. But during a certain period there was an extensive abuse of psychology by making it strictly mechanical. It was a religious scruple (because it is a religion) to describe behavior solely in terms of tropisms and reflexes. People spoke of physiology and never of psychology. The treatises written were more and more detailed, stuffed with pedantic expressions often beside the point. Who was it that maliciously described a certain psychological zone as the science which spends more and more time and more and more money to study more and more phenomena which are less and less interesting? Well, this is not altogether a jest. I will cite no country in particular though I am thinking of a country while speaking of this. I can refer to an apologue that I often cite to my students. It is the apologue of the Martian who finds a watch. Let us suppose that on Mars there is no watch (which is probable). How will he interpret it?

If he has not the slightest idea of this mechanism which he finds on the ground, he can practice anatomy. He can dissect the watch and analyze the order of its parts. He can note that the different dented wheels have not the same diameters nor the same number of teeth. This is useful. One can write long papers about it. If the Martian practices microscopic anatomy, he can study two different alloys of the watch and ascertain that those which constitute the wheels are not the same as those which constitute the case. This is not uninteresting. But, between ourselves, so long as the Martian has not understood that the watch is a machine to measure time (I do not say "a machine *made to* measure time," for this would be an imprudent statement) he will not have understood anything important and he will not be able to interpret the watch.

In addition, if he puts all the works in a paunch or a box specially designed for this purpose, he will have a perfect description of the watch but an essential thing will be missing. He will no longer be able to measure time. And yet, he will not have done anything wrong in itself by placing aside the pieces of the watch. It is in the interpretation, in the attitude, of the observer that there will be something profoundly inaccurate.

Let us return to our animals. If we wish to be coldly objective and to disregard the fact that we are men and a part of the same world, we find ourselves, when confronted by the mysterious biological world, somewhat in the same situation as a Martian in front of a watch. What are all these mysterious organisms swarming around us?

Let us take an example with which I am familiar. If when confronted by a beaver you have not understood that it is essentially an animal that builds dams, I do not know whether you have understood anything very important about the beaver even though you may have studied its physical and mental biochemistry. Nothing prevents you from studying the conditioned reflexes of the beaver. But will something fundamentally important concerning the beaver not have escaped you?

I do not say that the beaver is a machine *to* build dams but, sticking to reality more closely, I say that it is an animal *which* builds dams. This is a slight verbal hypocrisy but I italicize it. The fact that it is an animal which builds dams distinguishes the beaver from other animals amongst other things, and that is the important thing to study. The ways and means by which it builds dams is what must be studied and these ways and means are very complicated. Similarly, a panther is a mechanism which catches its prey. And what is a mole? It is a mechanism which digs the earth.

I seem to talk in truisms. As a matter of fact they are not really truisms. These elementary truths were neglected for a long time. Why? Because of the fear of being a finalist.

Evidently the example given by Bernardin de St. Pierre—namely that a melon has ridges so it can be eaten by a family—is farfetched. But it is also a little childish to admit that nothing is of any use, as did Rabot, a biologist who was ingenious in his youth. I remember reading a chapter by Rabot which was, in fact, charming. Rabot had a peculiarity: he was extremely upset by the eye. You will say that there was no reason for this and that the eye is unquestionably an optical instrument. But you will forgive me if I say that the eye is *made to* see and the ear is *made to* hear; or, more exactly, that the eye is a machine made to see and the ear is a machine made to hear. Rabot admitted this but said, "It is a wretched optical instrument. Let us leave it aside." This refusal to admit a part of reality is ridiculous. We must keep ourselves from excesses in one direction or in the other. When considering a determined organism, it is rather difficult to keep oneself from the two contrary excesses—an excess of finalism and an excess of mechanism. But it is impossible to ignore these two attitudes. One cannot forego a mechanistic analysis of behavior because it is useful and pays off, and one cannot forego a finalistic attitude in the very limited sense in which I use this expression. It would be absurd to do so.

A big question evidently remains. Though I do not wish to encroach on Professor Grassé's lecture, I would like to speak of the genesis of organic machines, because the other argument is evidently neo-Darwinian.

The neo-Darwinians take a very convenient stand, and I mistrust stands that are too convenient. For them the world has no mysteries. They know all there is to be known about the complete genesis of all organisms. I assure you that I do not exaggerate. I have just reread a passage by Julian Huxley, a man I greatly admire, but in my opinion, he should blush for the gross blunders he left at the bottoms of certain pages.

The very fact of evolution is not actually denied by any reasonable biologist. It is evident that the species have changed and that they have changed in different directions, in the direction of progress (if we can define progress, which I doubt) or in the direction of a regression (perhaps easier to define). In short they have changed, and they very certainly derive one from the other. This being said, what is the mechanism?

Some people, of whom I am one, think that the mechanism is still very obscure. Others, as I have pointed out, say that it is actually devoid of mystery. It is the play of blind mutations which through natural selection have ended by building, in many millions of years, the most complicated mechanisms. This would demonstrate that the finality of the marvelous organic machines is only an appearance. At least that is what the promoters of this thesis say, but their arguments may not be as solid as they think.

On the other hand, let us try to give a few precise details about these theories, for they are beautiful, deep, and ingenious. The trouble is that they are perhaps not true. They contend that the hereditary potential of living matter is modified by chance through mutations which are, moreover, very rare. Everyone has heard of this. One of these mutations sometimes happens to present an advantage. No matter how slight that is for an individual, that individual will have supplementary chances of survival and chances of transmitting this advantageous mutation. Everyone has read this in magazine articles. Thus at the expenditure of many millions of years the most complicated mechanisms have been built.

More recently the theory has been reformulated. Though it is not easy to know what some neo-Darwinians think when they are pushed to the limit, it seems that for certain authors, selection would only be an advantage of reproduction, an ability to bear more young or slightly more viable young. If the mutation which confers this advantage of reproduction is accompanied by another morphological or physiological particularity, this last will have greater chances of being transmitted. This is all right as far as I am concerned, but in this form the theory is nothing but a truism.

Then Fisher and several others, including Reicht, calculated that the order of frequency of the mutations was about comparable with the order of magnitude of the geological periods which can be fairly well known by fairly good methods. Following Fisher, if a supposedly dominant mutation gives a selection advantage of 10 percent, and the bearers of this mutation have one supplementary chance out of a thousand of being reproduced, we can say that for certain species this mutation will take around five generations to establish itself in the entire species.

This about tallies with what is observed in general paleontology for the different species of *Equidae*. I leave the responsibility for this affirmation to Fisher. This could explain the multiple derivations of the anatomy of horses since the "luctus" and of all those animals which resembled horses in a more or less distant fashion and who ended by giving birth to the genus *Equus*. Nevertheless several objections can be made to these calculations.

First, a theoretical objection: namely that mathematics will never give back more than you have put in. In other words, it cannot create new parameters, when you have arbitrarily introduced certain parameters. The thing that is troublesome in these calculations is the impossibility of making a crucial experiment, the impossibility of knowing whether these calculations are correct or false. After all they may be correct, but I do not know. At the best we remain in a realm of vague probability.

Then there is the whimsical choice of the character of the parameters. Certain of these criticisms are borrowed from Ruyer who has written several books on finalism.

Let us admit that there is some kind of selective advantage. We will suppose that it will remain more or less constant during 50, 100, or 200 million years. I am not averse to this, but it is a dangerous assertion. In any case you will never be able to prove that this is what happened. Quite the opposite could have happened. A fastidious biologist, who would choose his parameters otherwise, could prove the contrary without there being a greater possibility to disagree or agree with him.

It is the eternal question to which Professor Grassé will come back, I am sure. Certain adaptations or certain mechanisms are so fabulously complicated that it is not impossible but very, very improbable (there is an important difference) that they could have been constituted by the blind play of mutations, each of which would only bring a very trifling selective advantage.

In this connection I consider very objectionable a sentence which escaped Fisher in his *Evolution as a Process*. Fisher wrote that selective advantage could by definition be sufficiently slight to escape, again by definition, all our means of measurement. An advantage which by definition cannot be measured! What is this science by which arguments are employed that I have encountered in no other science? Biologists and scientists in general do not like theories which, hypothetically, cannot be verified.

Finally, the last objection (that of the eye and the ear) has not stopped certain neo-Darwinians who are the most imaginative of men. It is too bad that their imagination sometimes tends to be extravagant. They say that it is possible for certain mutations to act on more than one character; that is true. There is what is called the pleiotropic effect of genes. It can happen that one gene acts on one character. I think that any mutation acts, step by step, on the whole organism. This seems probable and even certain. The division that we establish between such and such a gene is in part real but also in part a result of our mental structures. It is, therefore, conceivable that certain genes could bring about in one fell swoop a very important effect of organization or reorganization and that a few mutations of this type, instead of an infinite number of mutations, would suffice to answer the preceding theoretical objection. From this point of view, I will not take as an example the eye and ear, for these examples are too complicated. I will take simpler ones.

There is a certain butterfly (*Papilio polytes*) some kinds of which can appear under several different aspects. Some females have commonplace coloring. Others are mimetic and imitate very precisely other noncomestible butterflies. The degree of imitation is sometimes surprising by its exactitude.

Mimetism is an amazing phenomenon; it is one of the thorns of biology. This applies to imitations by comestible species of noncomestible species that are astounding by their exactitude. When certain noncomestible butterflies are introduced into the beak of a bird or the mouth of a lizard, the bird or the lizard rejects them. They then rub their beaks or mouths on the ground as if they wanted to rid themselves of a disagreeable taste.

On the other hand there are comestible species which imitate with precision noncomestible species. If they are reduced to a ball and introduced into the beak of the predator, this accepts it perfectly.

I ask you to believe that it is not a question of approximate imitations. It is fantastic! Occasionally it deceives biologists who have sometimes put into the same box cockroaches and coccinella which resembled each other so much that the confusion was possible.

Certain imitations appearing in the *Papilio polytes* were analyzed by Ford. In this precise case it seems that the mimetic/nonmimetic change depends on only two modifying genes which shunt the organism in one direction or in another. This is a very rare phenomenon. I do not know whether there are any other cases where the genetic potential of a mimetic species has been analyzed with such great precision.

These mutations exist for they have been demonstrated. I could cite another case. I remember that long ago, when I worked with Professor Grassé, I had taken from one of his drawers a few samples of *Kallima* butterflies. Because of their resemblance to a dead leaf I had placed them on some dead leaves on my desk. When I say "resemblance" do not take my words in a restricted sense. The imitation is bewildering. The color is there, the veins are there, the leaf-stem is there, but furthermore there is the denticulation on the edge of the wing of the butterfly which imitates the irregularities of the dead leaves. Small transparent areas imitate the attacks made by the *aminus bacillus* in the parenchyma of the leaves which produce a kind of mirror in the interior of the leaf; there are spots that resemble moulds; they are butterfly scales, but these scales imitate the moulds with such precision that Roger Heim, professor of mycology, had to determine which were the imitation moulds. The whole affair is maddening.

Let us admit, for it is possible, that these extraordinary structural characters of the butterfly depend on a few genes. I am willing to agree. But how shall we interpret such a mechanism? We will simply have pushed the problem back. At that moment, the gene or genes will only be a wonderful deus ex machina. It will be a true miracle. It will not help to say that a gene produces this extraordinary imitation, either of another animal or of a dead leaf, by itself alone. I will always ask how that gene acts. The real problem is the *how*.

I must strongly insist that the thing which naturally exasperates the mechanists is the fact that there is no explanation. They say, "You are Lamarckians, or you have no other explanation." No, I have none. I simply propose to search. This world is very vast, very mysterious and in great part unfathomed, and I do not pretend to know everything about it. I have no ready-made solution. I think—and moreover regret it—that Darwinism, neo-Darwinism, and in great part the old forms of Lamarckianism, are nothing more than the childhood hypotheses of biology.

Many biologists, moreover, have noticed that neo-Darwinism proposes a very curious example of the intrinsic necessity of finalism. Indeed this principle of selective advantage, utilized with a certain imprudence by a great number of biologists, is nothing else but their old enemy, finalism, introduced through a subtle though evident slant at the beginning of things, whereas a certain number of philosophers introduced it at the end. But what is the difference? The problem is the same. It belongs to the same type of explanation that we encounter so often; the soporific property of opium. There is in living matter an organizing virtue whose characteristic is always to organize the greatest possible amount of utility.

This is a particularly disappointing aspect of finalism and particularly unfruitful, for one can get nothing out of it. In reality, how is such an explanation superior to the naively finalistic story of Bernardin de St. Pierre according to whom the melon has partitions so that it can be eaten in common? I wonder.

You could evidently tell me that all I am saying here is but a work of demolition, very easily done. As everyone knows, criticism is easy but art is difficult. And we could retort: what other solution do you propose?

I do not ignore the superficial aspects of all the criticisms I have made. But do me the justice to admit that in such a short time I could not be very profound. Moreover, this does not affect me for a serious reason; namely, that nothing equals my astonishment at the attitude of certain biologists who are confronted by the living world. As a biologist and also as a physiologist (though I have committed many infidelities to physiology, I was a physiologist in the past) I know as well as anyone to what a degree the animal or human organism is complex and to what a degree its mechanisms are little known. This is a first truth.

We have just made an important step forward through the discoveries which rightly obtained the Nobel Prize for three Frenchmen. It is a first step in a long obscure corridor, but it is nothing more. I am convinced that these three scientists, who shine by reason of their modesty, would not fail to agree with this.

Everybody admits that we ignore the constitution of an organism and yet pretend to know the genesis. To be kind we will admit that biology is actually at the initial stage of physics though certain scientists think it far from having reached this stage. How then can you expect modern biology to unravel the genesis of an organism of which you do not know the constitution? It seems to me that there is here an extraordinary aberration. It is very improbable that the secret of the universe—for this is neither more nor less than the secret of the formation of the organisms—should have been revealed all at once by a benevolent deity to certain scientists particularly favored by the gods.

I agree with Ruyer: living beings have a history; but this History (written with a capital *H*) is crossed by a thousand misfortunes, a thousand backslidings caused by different mechanisms among which it is very probable that natural selection plays a part, as well as, moreover, a hundred or a thousand other elements, a hundred other distinct processes. Why not?

We will some day perhaps guess what these processes are. I think that this universe is unfathomed but not unfathomable. I seriously believe that we will discover them in ten years, in a hundred years, or in ten centuries. At that time we will perhaps be able to propose a less immature idea of the evolution of organisms than the theories we actually have now. They can only be immature. I think that for the moment we can only seize two guides, one as lame as the other. On one side the mechanistic analysis of the processes. We cannot do without this. We analyze how a certain function gives rise to such or such a process and we can do nothing else from a strictly mechanistic point of view. On the other hand let us not be so naive as to veil our eyes to the facts that are in front of us and which are complicated mechanisms. They are really mechanisms; they really function; they are fantastically complicated and man, moreover, sometimes finds it useful to imitate them.

To seek to abandon this modest and restrictive position seems to me an untenable position, manifesting an attitude of blindness before the immense, the unfathomed, the mysterious universe.

The Role of Finality in a Philosophical Cosmology

Jean Ladrière

The idea of finality is a very old philosophical idea, but one which is far from being naive. However, the success of mathematical and mechanistic representations of nature has, for a long time, somehow intimidated those inclined to think in such terms. Throughout this century, prominent scientific thinkers—and among them, most particularly, the man whose memory we are honoring in this conference—have on the basis of purely scientific considerations, reintroduced the idea of finality and have emphasized its importance in an overall interpretation of nature. This fact puts philosophy under a kind of obligation to return to this problem and to try to recover the meaning of a very old insight.

Natural science presents us with two general categories of descriptions. There are descriptions in terms of process and descriptions in terms of structure. Of course the two are closely related, and there is a general tendency to reduce all descriptions to descriptions of the first type. However, for our purposes, it can be useful to consider the two types of analysis separately.

THE DESCRIPTION OF NATURE IN TERMS OF PROCESS

When we describe nature under the aspect of process, we use the general model of connections between states. It is well known that a precise definition of the notion of *state* presents great difficulties. Classical mechanics developed a simple definition in terms of position and momentum. For this kind of description, the state of a system is characterized directly by means of observable quantities. The same kind of approach is found in thermodynamics. But even in classical statistical mechanics it was necessary to introduce a more abstract definition—a definition which characterized the state of a system by a set of values which were observable in principle but not in practice. The quantities actually measured were related to these values through the probabilities of the various possible states. In quantum mechanics the state can no longer be defined in terms of observable quantities. Instead, it is characterized in an abstract fashion by a certain function which does not have a direct empirical reference. This function merely describes a field of possibilities. Since it is a distribution function, it expresses a property of an abstract space, which represents only possible dispositions of the real system under study. The physical quantities which describe this system are themselves represented in an abstract fashion; and it is only by applying the mathematical theory of eigenvalues that we return to observable elements (i.e., to the possible results of measurements).

However, whatever may be the difficulties connected with the notion of state, the fundamental problem is to find a way to describe connections between states as a function of time. We know how physics solved this problem—by means of (in general) second-order differential equations. We suppose that the development of a physical system satisfies certain conditions of continuity which imply that the functions which describe the system and its interactions must be at least twice-differentiable. The equations by which we represent the process of this development involve differentials; that is, they characterize the way that a global process occurs by describing an infinitesimal process. Such a description allows us to characterize the course of a global process in such a way that it is possible to reconstruct this course in terms of the infinitesimal description. This is precisely the purpose of the operation of integration. This operation introduces indeterminate factors in the form of the constants of integration. But if we know the state of the system at any given instant, we can remove the indeterminate factors. Then by an algebraic calculation we can determine the state at any prior or subsequent instant. Thus, the solution of the problem posed leads to the establishment of a formalism which yields predictions and retrodictions.

Depending on the nature of the system under study, we must employ different types of description. In other words, in physics we need a multiplicity of laws of connection. But the point to be emphasized is that the construction of all these laws is carried out in accordance with a general principle. This is the principle of the connection of states, which can be formulated as follows: The states of a physical system are connected to one another in an unambiguous way, so that it is always possible to determine, on the basis of a given state of the system (which state we suppose to be described completely) all the past and future states of the system. The restriction introduced concerning the completeness of the description is essential. A complete description is possible only for a system which is permanently closed. If we are considering a system which is closed for only a certain period of time, we must, in order to give a complete description of this system, determine its place in the larger system of which it is a part. This means giving a description of the larger system. And this description itself will be complete only if this larger system can be considered permanently closed.

The principle of the connection of states, as we have formulated it, is valid in all its generality only for permanently closed systems. As Grünbaum has shown, there are situations in which the symmetry between prediction and retrodiction does not hold.[1] There are cases where knowledge of the state of a system at an instant t_0 allows us to determine the state of the system at an instant prior to t_0 but not at an instant following t_0. Likewise, there are cases where knowledge of the state of the system at an instant t_0 allows us to determine the state of the system at an instant following t_0 but not at an instant prior to t_0. The first kind of situation occurs when the system under study has been momentarily isolated from a larger system and is in a state of relatively low entropy. Such a system presents a configuration which can be interpreted as an indication of a prior interaction. But the configuration does not allow us to make any predictions about future interactions of the system. A prediction would be possible only in the framework of a much larger system which contained both the given system and perhaps numerous other partial systems. The second kind of situation occurs when we are dealing with an irreversible process, such as a phenomenon of diffusion in a momentarily isolated system. For example, if the system is in a state of disequilibrium at a given instant t_0, we can predict the future states of the system, but we cannot reconstruct its past on the sole basis of information available at the instant t_0. The state of disequilibrium may have been produced by a prior diffusion, but it also may have been produced by an interaction with an outside system. Here a retrodiction will be possible only if we have access to more general information; i.e., information about a more extended system which contains the given system.

We must distinguish between strong connection and weak connection. Strong connection is that of classical determinism. It is possible to predict and retrodict the states of a system with any degree of accuracy desired. Of course, this presupposes that the state of a system at any given moment can be known with any degree of accuracy desired. As several authors have emphasized, such a presupposition represents a simple idealization of experience. Actually, even from a classical viewpoint, there is a limit to the precision of our observations, since we must take account of the thermomolecular movement and this has a random character. When we try to obtain measurements that are more and more precise, we necessarily reach a point where the order of magnitude of the measurements made corresponds exactly to the order of magnitude of the random movements. Classical physics takes account of this fact, since it employs a theory of observational errors; but it covers up the fact (so to speak) because this theory of observational errors is separated from the predictive formalism, which in itself is regarded as dealing with absolutely precise values.

Accordingly, even in classical physics we must distinguish between strong connection and weak connection. There is weak connection when there is a limit to the precision of possible predictions and retrodictions, whether this is because the present state cannot be known with certitude or because the very nature of the connection involves a completely random element. In classical physics there is strong connection in principle but weak connection in actual practice.

Quantum physics can be said to ratify actual practice, since its predictive formalism applies no longer to ideal values, as was the case with classical mechanics, but to the possible results of measurements that can actually be carried out. We might say that quantum physics incorporates its theory of observational errors into its predictive formalism. This incorporation is signaled by the occurrence of probabilistic language in the predictive formalism: the laws of quantum mechanics are statistical laws. Accordingly, the connection is no longer unambiguous. It deals with probability distributions. The result is that if we still employ the classical description in terms of position and momentum, we can predict only the probability of observing a possible value of the physical quantities under consideration. But a redefinition of the notion of state allows us to return to the model of strong connection. Quantum dynamics in principle allows us to determine precisely and unambiguously the wave-function at any instant if we know this function at a given instant. If we regard the state of the system as determined by the wave-function, we then recover an unambiguous connection between states. But of course this requires us to give up spatio-temporal description and adopt a less intuitive description. This description is given in terms of an abstract space-configuration space, which is, after all, simply a set of possibilities—and not in terms of real physical space. However, this description still has real significance, since knowing the state function allows us to determine the possible values that can result from actual measurements of the physical quantities that characterize the system under study.

We can ask ourselves to what extent such a predictive formalism is in accord with the classical notion of causality. Let us consider the principle of causality as it was formulated by Kant. In the first edition of the *Critique of Pure Reason* we read: "Whatever follows or happens, must follow in conformity with a universal rule upon that which was contained in the foregoing state."[2] And, in the second edition, "All changes take place according to the law of the connection of cause and effect."[3] More simply, we can take Hume's formation: "Like objects, placed in like circumstances, will always produce like effects."[4]

As Lindsay and Margenau have pointed out, the principle of causality understood in this manner is not identical with the principle of connection which we have been discussing. For we can propose the following line of reasoning. Imagine a universe controlled by a capricious demon who acts solely by whim. Such a universe would certainly be considered acausal. However, if the movements imparted to the universe by our demon are not too discontinuous and if changes arise from relatively small variations, we can introduce convenient hypotheses of continuity into our description of this universe and analyze it in terms of differential equations. This will provide us with a formalism expressing connections between states. Thus, it is quite possible to have connection (in the sense, indicated above, of predictability and retrodictability) without causality. According to Margenau and Lindsay, causality (insofar as the term can be given a meaning in an empiricist theory) signifies regularity in the dependence of one state on another; or, in other words, invariance with respect to the time of the functions which describe connections between states. More precisely, these authors propose the following definition of causality in the form of a principle of the consistency of nature: "The differential equations in terms of which nature is described do not contain explicit functions of the time."[5] Here we have a translation of Hume's formula into the language of physics. It is clear that the extended meaning of the principle of connection which is introduced by quantum physics is in no way incompatible with this kind of conception of causality. The fact that the connections are described by statistical rather than by deterministic laws does not of itself introduce any irregularity into the universe. On the other hand, the presence of causality would be incompatible with the complete absence of any sort of connection. In other words, where there is causality, there is also connection. Thus, the principle of causality, as we have defined it, is a less general principle than the principle of connection. Causality implies connection, but connection does not imply causality. In other words, causality is a sufficient but not a necessary condition for connection.

What must attract our attention is the fact that these two principles, the principle of causality and the principle of connection, are regulative principles. They do not refer directly to phenomena but to the form that must be taken by the equations which describe phenomena. In this sense, they have an a priori character and, at the same time, a unifying function. They indicate in advance and independently of experience the way that laws must be formulated and are, therefore, a priori. Further, they do not as such refer directly to the content of experience. It is only indirectly, by means of the consequences that can be logically deduced from them, that they can be connected with this content. On the other hand, they unify diverse laws by indicating the general model in terms of which laws must be formulated. Through the intermediary of particular laws, they refer to the totality of phenomena. Moreover, they can do this only insofar as they are a priori. This means that in these principles we have a global understanding of nature which precedes and controls our understanding of details. The true reason for the intelligibility possessed by particular laws is due not to their applicability to experience but to their conformity to a universal principle. In other words, the understanding of the parts is possible only on the basis of an understanding of the whole. We cannot regard the understanding of the whole as a kind of limit-idea, as if we could only approach it indefinitely by putting together, one after the other, the partial knowledges that we are able to obtain about nature. On the contrary, it is partial knowledge which is a limit-idea. It is by basing ourselves on an understanding of the whole that we approach more and more the knowledge of details. Naturally, this understanding of the whole is not, properly speaking, really knowledge: that is, it does not put us in the presence of an object, but only indicates the general form that must be taken by any knowledge; i.e., the general way that objects may be given to us. Thus, it is an indication of a certain type of giving.

This indication corresponds exactly to a certain idea of nature which is presupposed by natural science. The directive principles, inasmuch as they indicate the general form of laws, are principles of natural science. But inasmuch as they correspond to an a priori understanding of the very object with which this science is concerned, they are the expression of a global grasp of nature given in the precise form of an indication of a type of giving (i.e., of a type of manifestation). In this sense, they express an idea of nature, a demarcation of the realm of occurrences which is nature, an ontology of nature. They are the expression of a prior ontological understanding of nature, an understanding which is presupposed by scientific knowing of nature. But they are only an expression of this understanding, not a reflection on it. This means that in scientific knowing the prior understanding which this knowing employs remains implicit.

Now physics, still considered from the point of view of the description of processes, gives us still other principles which have this regulative character. As examples we can cite variational principles and principles of conservation. As is well known, it is possible in mechanics to derive the equations of motion from a variational principle. Consider, for example, Hamilton's formulation of this principle. Suppose we have a conservative mechanical system—a system in which no energy is lost. For such a system we can define the kinetic energy, which depends on the positions and velocities of the particles which make up the system, and the potential energy, which depends solely on the positions of these particles. We form an expression L by taking the difference of the kinetic and the potential energy. This expression is called the kinetic potential. Next, we consider a motion of the system which occurs between t_0 and t_1. We integrate the expression L over time between the limits t_0 and t_1. In other words, we decompose the motion into infinitesimal elements. With each of these infinitesimal motions we associate an infinitesimal quantity obtained by taking the product of the kinetic potential and the infinitesimal length of time during which the infinitesimal motion occurs. We form the sum of all these quantities, thus obtaining a measure of the total action of the kinetic potential in the course of the motion considered. We can now take the variation of this integral and look for the conditions under which this variation is zero. In other words, we consider all the possible motions of our system which have the same initial states and the same final states. To each such motion there corresponds a value of our integral. We look for a privileged value which is an extremum; that is, which is either larger or smaller than all other values. Hamilton's principle is the following assertion: the actual motion of the system is that for which our integral has an extremal value. Depending on the nature of the problem, we will be dealing with either a minimum or a maximum value.

We can easily generalize this principle and extend it to the case of nonconservative systems (where energy is lost) by introducing external forces acting on the system. As is well known, the idea of variational principles has been successfully extended to nonclassical mechanics. Thus, in the general theory of relativity, we can derive the equation of motion for a particle in a gravitational field by integrating the infinitesimal element of distance and looking for the conditions under which this integral has an extremal value. In this way, we arrive at the relativistic generalization of the principle of inertia. A particle in a gravitational field follows the shortest path; it moves along a space-time geodesic.

A variational principle does not give a direct description of motion, the way a differential equation does. Rather, it indicates a general condition that the motion must satisfy; and this is expressed formally by the fact that we can derive from it differential equations which describe the motion. The differential equation treats the motion as a becoming, *in fieri;* it allows us to reconstruct the motion, by a kind of step-by-step procedure, in terms of its local behavior. The variational principle treats the movement globally and specifies a condition that any actually occurring motion must satisfy. Moreover, this condition is stated in terms of an extremum; to this extent it corresponds to what, in another language, has been called an optimum principle. In fact, we could rephrase the principle as follows: among all the paths that are theoretically possible, nature has chosen as the privileged path the one that is the simplest. Thus, in the case of general relativity, the particle always follows the shortest path, as if nature strove to attain its ends by the least devious and most economical means. However, the consideration of all possible paths is merely an artifice; in reality, only one path is really possible —the path, in fact, followed by the particle. It is only by abstraction that we can treat the motion of the particle as an isolated phenomenon. In reality, its motion is conditioned by the overall configuration of nature; it depends on all other motions.

We can easily see this in the case of the relativistic particle; the particle follows a geodesic, but geodesics are determined by the structure of space-time; and this structure itself expresses the configuration of the gravitational field. Further, this latter configuration depends on the distribution of masses in the universe, and this distribution depends on the motion of the masses. Thus, the variational principle expresses the way the motion of a particular particle under consideration is conditioned by the set of all natural processes. Therefore, it expresses a connection between a part and the whole. More precisely, it prescribes the form that this connection must take. It tells us that every particular motion must be carried out in accordance with the rule prescribed by this form. Here again we find a priority of understanding the whole with respect to knowledge of the parts; it is only on the basis of a certain understanding of the whole that the analysis of particular phenomena can be accomplished. And in the understanding of the whole, there is expressed a certain idea of nature. Here we add to the general notion of systematic connection (or stability) the notion of simplicity in performance; and this notion can be regarded as a condition of harmony. If the principle of connection is the condition for systematic knowledge, the variational principle is the condition for an aesthetic perception of nature.

Let us now turn our attention to principles of conservation. Classical mechanics had already introduced two famous conservation principles, one concerning matter and the other energy. The special theory of relativity has allowed us to reduce these two principles to one. The successive extensions of the notion of energy in the course of the development of physics have been accompanied by corresponding successive extensions of the principle of the conservation of energy. The general conservation principle is today regarded as a fundamental principle of natural science. But modern physics has been led to introduce other fundamental quantities and associated principles of their conservation. Since it is possible to link these conservation principles to invariances under certain groups of transformations, one might think that in principle it must be possible to unify them within a broad general theory of imbalance. In fact, such a theory is presently almost a reality. What we must emphasize here is that these principles play a fundamental role in understanding interactions between elementary systems. The only interactions admitted are those which obey the various conservation principles. To put it another way, the interactions must occur in a way that leaves certain expressions invariant.

Once more, we find here a general principle of a regulative character which plays the role of a unifying a priori for the particular laws which describe interactions. To a certain extent, this principle of invariance recalls the principle of causality, since it asserts a basic stability in nature. However, it seems to be more general, since it does not concern only invariance under translations along the time axis and since, on the other hand, the principle of causality, in asserting temporal independence, expresses only what we might call a trivial invariance. Likewise, the principle of invariance, to a certain extent, recalls variational principles, since it expresses a kind of economy of means in natural processes. The transformations of nature must occur in a way that changes the configuration of nature as little as possible. In other words, only certain motions are allowed. And this condition is certainly a condition of compatibility: the motions which can actually occur are those which are compatible with all others; that is with the overall configuration of the universe (taken simultaneously in its spatial and its temporal extension).

However, there is a difference. Variational principles pertain to motions in space-time, whereas conservation principles have a more general form, since they apply not only to spatiotemporal displacements but also to transformations of configurations (for example, the substitution of one elementary particle for another in the composition of a physical system). Of course, a synthesis may be possible. If we can succeed in expressing everything in terms of a single field, we may be able to explain conservation principles on the basis of the very general conditions imposed on this universal field; and perhaps these conditions could be expressed in the form of a variational principle. In any case, here again we find a determination of particular laws by reference to the totality. For, in the statement of conservation principles, we express an idea of nature as a totality. And it is precisely because they express such an idea that these principles can play their regulative roles.

THE DESCRIPTION OF NATURE IN TERMS OF CONFIGURATIONS

So far we have considered natural science under only one aspect: that of descriptions in terms of processes. But our discussion of interactions has already obliged us to consider another aspect of natural science: that of architectonic descriptions—the analysis of configuration. Indeed, experience shows us that natural objects present themselves as figures that are more or less stable (at least if we consider sufficiently short periods of time) and relatively individualized. Moreover, these figures are arranged in a hierarchy. To each rank in the hierarchy there corresponds a certain type of organization. And each level functions as a necessary (but perhaps not sufficient) condition for the occurrence of the next higher level. The projection that natural science must adopt with respect to this aspect of the universe is, first of all, to characterize the different levels of organization and then, if possible, to explain them in terms of processes, by means of the fundamental forces that appear in the analysis of processes. This is one version of a universal program of reduction; we must try to explain a given level of organization by reducing it to the level preceding it. We must try to show that interactions which take place on the preceding level can, given appropriate conditions, produce the occurrence of the configurations of the level in question. If we had succeeded in developing such an explanation for all levels, then we could reduce, step by step, any material phenomenon at all to an elementary level. In the present state of our knowledge, this elementary level seems to be that of what we call elementary particles. But it is possible that we will endow these particles with a structure or consider them as themselves being the results of interactions occurring on a lower level. It is even possible that we will someday succeed in describing everything in terms of a single field. Then we could give an account of all the world's phenomena on the basis of the laws of this universal field and thus reduce every explanation to the action of some very general principles of invariance. But the great question which arises is that of emergence. How are we to understand the passage from one level to the level immediately above it?

Three models of explanation seem to be possible: (1) the occurrence of the new type of configuration is a necessary consequence of conditions on the lower level; (2) the occurrence is a chance event; (3) the occurrence must be explained simultaneously by conditions belonging to the lower level and by conditions not belonging to any lower level.

In the first case, it is simply a matter of applying the model of connection. The phenomenon of emergence is completely superficial. Actually, the new configuration is just the result of prior conditions. And the phenomenon of emergence is explained by the fact that we are dealing with a complex system which is made up of a large number of subsystems. Here we can take statistical mechanics as an example. If we consider a system composed of a large number of particles, we will note that the system displays properties, like temperature, which cannot belong to the particles but which can, nonetheless, be explained by reference to properties of the particles. Conditions realized on the lower level are thus at the same time necessary and sufficient conditions for the phenomena of the higher level.

The second case can be divided into three subcases. That is, we can understand the intervention of chance in three different ways: (a) it may be due to a gap in the information which we have about the lower level; (b) it may be due to a complete lack of causality; (c) it may be due to conditions that are not present on the lower levels. In the first subcase, we regard the new configuration as just one of many possible configurations. All these possible configurations can be explained on the basis of conditions present on the lower level. But for this case the connection is ambiguous; and the result is that, taken by themselves, the conditions given on the lower level do not suffice to determine the configuration that actually occurs.

However, we grant that there are on the lower level additional conditions that would allow us to explain why such a configuration occurred rather than any other. The appeal to chance merely expresses the incompleteness of our information. Another possibility is that as time passes, the processes which unfold on the lower level (although they generally yield unstable configurations) can at a given moment yield a stable configuration. As before, we grant that there are unknown conditions which would allow us to explain why a stable configuration occurred at that given moment. And again as before, chance corresponds simply to a gap in our information. In other words, if there seems to be a chance event, this is merely because we have not succeeded in closing the system. If we could do this—if we could obtain a complete description of the lower level —we would again find an unambiguous connection.

In the second subcase, we grant that it is not possible to complete the description and that, consequently, we encounter an indetermination that cannot be eliminated as long as we remain on the lower level. In this case we appeal to chance in order to assert that there are completely random events (in the strong sense of "random," i.e., "uncaused")—events which can in no way be inferred from antecedent conditions. In the third subcase, we again grant that the description of the lower level cannot be completed and that we must have recourse to other conditions; but since we have no way of representing these other conditions, we appeal to chance to express our ignorance.

In the third case, finally, either we are content to assert that in principle there are some conditions somewhere which, in conjunction with those present on the lower level, do explain the occurrence of the new configuration; or we try to specify the general nature of these conditions or even to identify them. But the problem is to give a physical meaning to such conditions. Here the modern theory of information provides invaluable assistance by (as Costa de Beauregard has so brilliantly shown) allowing us to regard these additional conditions as certain types of information. Since it has been proved possible to give the concept of information a physical meaning and to relate it to the traditional concept of negative entropy, it is in principle possible—even from a strictly scientific viewpoint—to appeal to a model of explanation which introduces emergence in the strict sense; i.e., which does not consider the conditions present on the lower level as sufficient and which, on the other hand, does not accept the idea of objective chance. It seems that this kind of explanation consists in showing the presence of a connection which corresponds to the possibility of a retrodiction but not of a prediction—

a connection which permits inference from a subsequent to a prior state but not conversely.

However, our appeal to the notion of information does allow us to regard the information as already present in the universe when the process which leads to the new form begins. Then the process becomes nothing other than the application of the available information to a preexisting material; we can compare it to the carrying out of a program. Thus, in a certain sense, we come back to the first kind of explanation. By adding a certain amount of information to the conditions present on the lower level, we do, after all, complete the system. An adequate description is given, and the connection between the lower level state and the resultant state (the occurrence of the new form) then becomes a perfectly determined and unambiguous connection. However, the problem is to locate the available information. If we take advantage of the equivalence between information and negative entropy, as Costa de Beauregard has done, we can indeed say that the information is produced at the expense of negative entropy. The increase in organization that occurs at one place in space-time is balanced by a loss of organization at another place in space-time. To be able to completely reduce the third type of explanation to the first, we must know the laws governing the transformation of negative entropy into formation. We could, of course, assume that negative entropy is information. In that case the preceding problem disappears, but then we must look for the laws that govern the passage of information from one system to another. If we follow this approach, we will probably be led to link the problem of emergence to the cosmological problem. Indeed, we will have to appeal more and more to earlier states of the universe and, finally, to an initial state which must contain all the information needed to explain every aspect assumed by the universe in the course of time.

This is precisely the kind of approach provided by Lemaître's model of the primeval atom. This primeval atom must indeed be regarded not only as an initial state in which the radius of the universe has a minimal value but also as an initial state of maximal negative entropy. It represents not only a geometric initial condition but also a thermodynamic one. For this theory, the expansion of the universe is accompanied by a progressive fragmentation of the primeval matter; and this fragmentation is itself a process of increasing entropy. The entropy of a complex system can be taken as proportional to the sum of the probabilities of the macroscopic states compatible with its given total energy, and the probability of a macroscopic state is proportional to the number of microscopic states which correspond to it. Therefore, we see that a system which can be in just one simple state will be a system of maximum negative entropy and that the entropy increases with: (a) the number of constituents; (b) the number of possible energy states of the constituents; (c) the degree of degeneracy of these energy states. (These relations hold because it is the degree of degeneracy which yields the probability of a microscopic state and because the probability of a macroscopic state is obtained—at least from a classical point of view—by multiplying the number of microscopic states which yield it by the common probability of these states.)[6] Consequently we foresee the possibility of giving a physical explanation of emergent phenomena in terms of an absolutely first initial condition. But here we are dealing with an explanation which is not of the first type, since it does not consist in an appeal to a process but to a pure given.

Perhaps we can draw our conclusions more neatly by examining a particularly clear example, an example for which the analysis of emergence has been completely carried out. This is the case of atomic structures. The problem here is to explain the different types of arrangements of electrons around a nucleus, arrangements which correspond to the known atom. As in all such cases, the problem is to understand why we encounter stable configurations formed from elements belonging to a lower level of organization. Here the Pauli exclusion principle is introduced. This principle has a completely general validity and allows us to determine the possible states of a complex system formed from elementary systems of electrically charged particles of spin one-half. The principle tells us that the only states that can occur are those which are antisymmetric under an interchange of coordinates between two elementary systems. It follows from this principle that no two electrically charged elementary particles can occupy the same state; and this conclusion is the guiding principle for explaining atomic structures and their regularities. The general problem that nature must solve is that of neutralizing nuclei which have charges equal to integral multiples of a certain unit charge.

This neutralization is brought about by the addition of an integral number of electrons. Electrons are added one by one in a way which satisfies the following two conditions: (1) each new electron occupies the lowest possible energy level available; (2) the exclusion principle is not violated. The possible states are themselves described by sets of quantum numbers which have integral or half-integral values. It is apparent that the possibility of a complex arrangement like that of the atom is already present in certain properties of the lower level; namely, in this case, the quantum states that can be occupied by the electrons. But there is a new element which intervenes, a law governing the arrangement of the elementary constituents. For the present case, this is the exclusion principle. And this principle is by no means a global property which can be expressed in terms of the properties of parts, as is the case in statistical mechanics. Rather, it is a regulative principle which has the status of information. This information takes on a material form for us in the configurations which actually occur. But it must, in some form or another, exist before these configurations in order to be realized in them.

The same model applies on the other levels of organization. The conditions of possibility for the new configuration must be already present on the lower level. Already on the level of the constituents, we must find a structure which is already sufficiently diversified and complex to be able to carry certain information. But on the other hand we must add an extrinsic condition (further information) in order to explain the actual occurrence of a new configuration. Since the structure present on the lower level is itself already a material expression of information, we are dealing with a hierarchy of information—a hierarchy in which the information of a given level exercises a certain control over information already present on the lower level. Thus we find in these structural considerations the same thing that appeared in our study of processes. It is finally the intervention of regulative principles that makes possible the construction of ever more complex configurations. Now principles of this kind (like the exclusion principle) express a condition relative to a certain totality: from the point of view of the constitutive elements, the emergence of a new form of organization means a passage into a totality. Regulative principles are totalizing principles. To the extent that we must, step by step, trace all information back to the initial state of the universe, we relate partial totalities to this totality as such, which is the absolutely primitive initial state.

Thus we see that there is indeed a priority of the whole over the parts. But we have encountered this priority in two different forms: on the one hand, in the form of regulative principles controlling the multiplicity of processes; and, on the other hand, in the form of organizing principles controlling the multiplicity of structures. In one case we have an order of law; in the other we have an immense initial condition, an order of fact. Even if we could explain all configurations in terms of processes, we would still very probably have to appeal to initial conditions.

In the principles which govern laws, there is expressed a certain pre-understanding of nature. In the principles which govern structures, there is expressed a certain original giving (a kind of pre-giving) of nature. In both cases we encounter a priority of the whole over the parts but this whole is never given to us as such. This is why the principles which represent the influence of the whole are only regulative principles. However, it is by reference to this whole that we are finally able to understand particular phenomena. To this extent, there is a certain kind of finality present in physical science; for the idea of an influence of the whole on the parts is precisely the idea of finality (at least in one sense of the term). Here we can quote Jules Lachelier: "To say that a complex phenomenon contains the reason for the simple phenomena which conspire to produce it is to say that it is their final cause."[7]

However, although finality is undoubtedly at work on the level of science, it is certainly not there as an object of reflection. This is why, properly speaking, the influence of the totality on its parts is never explicitly represented. For the case of processes, we have only an indication of the form of the laws, not a deduction of particular laws from a principle of totality. In the case of structures, we have only a description of types of configuration, and this is given in terms of emergence. We do not have a deduction of the properties of the lower level from more complex configuration. We can progressively reconstruct—and thereby explain—more complex configurations on the basis of given configurations; but we cannot grasp the general principle of all possible configurations because we do not know the initial state.

We might say that each step in scientific understanding is directed towards (and therefore implicitly contains) the totality, but that this totality is never represented. We might also say, in the language of Kant, that the totality is a concept for which there is no corresponding intuition; it is simply a regulative idea which lets us develop an architectonics of reason. By the same token, the initial state is as much a regulative idea as the totality of processes. By its very nature, it lies outside of all explicit representation; not because it is initial but because it is absolutely simple. Moreover, we can perhaps say that the two ideas merge into one. For the totality of processes, just as the laws of configuration and for the same reason, must be precontained in the initial state. To say that there is no reflection on finality, or that the totality cannot be represented, or that it is not given in an intuition—all these assertions seem to be equivalent.

TWO HISTORICAL MODELS OF REFLECTION ON FINALITY

But everything we have said is true only on the level of a knowing which is connected with intuition, a knowing whose proper intelligibility is based on what Kant calls construction in intuition—the operation of intuition. Physics understands nature by (so to speak) mimicking natural processes. We understand processes by manipulating the formalisms in terms of which we can describe them, and we understand structures by reconstructing them for ourselves. For example, we understand atomic structures and the periodic table of the elements by reconstructing them for ourselves by means of the rules of quantum theory. If we knew the rules governing the organization of the cell, we could understand it by reconstructing it abstractly, as in fact we are already beginning to do. By its nature, operational knowing cannot be reflective. But is it not possible to try to reflect on that which is given in this knowing, that which is represented in it under the form of operations? When we try to understand processes and structures, we understand them only in an implicit fashion, the way we understand a word in using it. Is it not possible to try to understand the process or the construction as act? Is it not possible to try to recover that which is exercised by the operation? To attain the operation as an act? Here we encounter the question posed by a philosophy of nature.

This philosophy cannot consist in synthesizing the diverse results of science or in employing some special kind of intuition which would not give us nature the way that science does. We have seen that the regulative principles of the science of processes bring into play an ontological pre-understanding of nature. Our problem is to transform this into an understanding properly speaking, to reflect it. In such a reflection, the totality will be thematized; but this does not mean that it will be given in intuition. Likewise, the notion of an initial state is only directed towards the totality as towards a universal potentiality. There is no question of knowing this state's content (which would amount to supposing an intuition of the whole), but rather of grounding its possibility (reflecting that towards which it is directed). In this way there arises the project of a reflective knowing which must bring to light the implicit presuppositions of science and, by so doing, exhibit the foundation of its possibility. But at the same time this knowing must produce an understanding of nature as passage, insofar as it is, in its entirety, one immense process. It is only in the framework of this kind of knowing that an explicit thinking about finality can be developed.

For finality is the action of the totality on the parts. Therefore, it is only by means of a knowing which strives to understand this action as such that a reflection on finality will eventually be possible. But what perspective can we adopt in order to lead ourselves towards such a knowing? History presents us with two models. If we wish to speak in terms of personalities, we can call them the Kantian model and the Hegelian model.

For Kant, as is well known, the philosophy of nature is articulated on two levels. On the one hand, it is a philosophy of physics; and, on the other, it is a reflection on nature as an idea of pure reason. Only physics gives us a true knowing of nature in the sense of letting us know its content. According to Kant, knowledge for us has content only insofar as there is a giving; and, for us, there is giving only in sensible intuition. Sensibility, in fact, is nothing else than the ability to be affected. The only intellectual factor in our knowledge has a formal nature: it is a priori (independent of experience); but this can be so only insofar as it is formal. Physics employs the a priori, but it does not thematize it. It is the task of philosophy to explain how physics is possible, or how a knowing in terms of the a priori is possible. This explanation necessarily reveals the foundation of physics. It is carried out in two steps.

In the first step, Kant analyzes the structure of the a priori. Here he sees that his a priori itself involves two levels—that of understanding and that of reason. The understanding is the faculty of rules, and the reason is the faculty of principles. Rules allow us to reduce diverse phenomena to a unity. Principles allow us to reduce diverse rules to a unity. The use of rules is, consequently, profoundly different from the use of principles. Rules are designed to be applied to what is diverse in sensibility; the concepts they employ correspond to intuitions and, consequently, the use of rules leads to the constitution of a true object of knowledge. Principles are designed to serve as guides for the employment of the concepts of the understanding; and since they do not correspond to any intuition, we fall into transcendental illusion when we treat them as if they gave us objects. The analysis of the rules of the understanding shows us the conditions under which a knowing of the world is possible. Since the conditions for the possibility of knowledge are at the same time the conditions for the possibility of the object, and since there is an object only in virtue of the employment of the concepts of the understanding (in conformity with the principles of pure understanding), it follows that the a priori structures of the understanding determine in advance the form that any possible experience of the world must take for us.

In a second step, Kant carries out, on the level of the a priori, a reflection on Newtonian physics and thereby founds its possibility and its validity. This reflection consists in showing how the pure concepts of the understanding can be applied to the given of motion. It turns out that the determination of motion by the categories (in conformity with the principles of pure understanding) leads to precisely the fundamental principles of Newtonian physics, which are thereby given an a priori foundation.

However, the pronouncements of physics do not exhaust that which can be said on the subject of nature. They correspond only to what Kant calls the determinant judgment, which is produced by the understanding. But in addition to this kind of judgment, there is the reflective judgment, which is produced by reason. Judgment in general is "the faculty of thinking the particular as contained in the universal."[8] But this can be done in two different ways, depending on whether we start from the general or from the particular.

> If the universal (the rule, principle, or law) is given, then the judgment which subsumes the particular under it is *determinant*. This is so even where such a judgment is transcendental and, as such, provides the conditions a priori in conformity with which alone subsumption under the universal can be effected. If, however, only the particular is given and the universal has to be found for it, then the judgment is simply reflective.[9]

It is precisely this reflective judgment which constitutes the judgment of finality. The empirical laws which science discovers undoubtedly conform to the a priori determinations of the understanding, but these are not sufficient to specify completely the form of these laws. Therefore, there remains in them an aspect of indetermination and of contingency. Since the supreme law of transcendental subjectivity is a law of unity, it is necessary to introduce a principle of unification which is distinct from the understanding. This principle, therefore, cannot correspond to an intuition and cannot serve to determine nature; it cannot enter into the constitution of a natural object. Thus, it is merely a maxim of judgment which "only represents the unique mode in which we must proceed in our reflection upon the objects of nature with a view to getting a thoroughly interconnected whole of experience."[10] Now such a principle must necessarily take the form of a principle of finality. Indeed, it has to be invoked because the understanding leaves nature partially undetermined. Therefore, it must consist in introducing the supplementary determinations necessary to give reasons for every contingency in nature. It can do this only in the fashion of an "as if" —by making the hypothesis of a total determination by an understanding, an understanding which obviously cannot be ours. Put another way, the principle consists in this assertion: the particular laws of nature must be regarded as if they conform to legislation enacted by an understanding capable of constituting a total system of experience. But such a principle is precisely a principle of finality. As Kant tells us, "The concept of an object, so far as it contains at the same time the ground of the actuality of this, is called its *end*."[11] To understand nature from the perspective of finality is to represent it in accordance with a law of unification, "as if an understanding contained the ground of the unity of the manifold of its empirical laws."[12]

How is such a representation possible? By virtue of the unifying power of pure reason. In the same way that the understanding unifies what is diverse and subsumes it under its categories, reason unifies the laws produced by the understanding by conceiving them in terms of the horizon of the unconditioned. Laws merely assert conditions and, consequently, indicate how chains of conditions are constituted. By reason we go beyond the domain of conditions by subjecting every series of conditions to a supreme conditioning principle—a principle which is not one object among others but rather the very condition for the appearance of all objects, the universal horizon of their constitution. Of course, this horizon is not given, and therefore we cannot constitute a knowing of it; we can think it, but we cannot represent it. But it is effectively present with all particular objects as the absolute system to which they all belong. In other words, reason makes nature appear as a totality. And we can understand the reflective judgment as laying down a law of totalization; it supplies what the human understanding lacks in order to give a reason for everything in the world that is contingent and particular. But it can do this only on the basis of a perspective of absolute unification in which nature is thought as nature. And this perspective is only a perspective of a horizon of constitution, not of intuition. Consequently, the further determination added to the unifying function of the understanding is purely formal; the judgment of finality cannot be determinative. "The highest formal unity, which is based upon ideas alone, is the unity of all things in accordance with an aim or purpose; and the speculative interest of reason renders it necessary to regard all order in the world as if it originated from the intention and design of a supreme reason."[13] Thus, we see that the Kantian philosophy of finality is foreign to physics, since physics rejects determination by an end and therefore knowledge of an end. Nevertheless, it is related to physics, since it assigns to the judgment of finality only the simple function of unifying the diversity of the particular laws of physics and thus of giving a foundation to that which cannot be founded solely by an appeal to the categories of the understanding.

If we now turn to Hegel, we find a philosophy of nature of a very different style. Instead of employing a knowing of nature as a point of departure for a thinking of nature, Hegel strives to think it at the very interior of knowing. But the knowing which is referred to here is no longer that of physics. Rather, it is knowing pure and simple —philosophy—a knowing which, in the course of its development, "salvages" physics by reinterpreting the givens which physics supplies within the infinitely more vast framework of absolute knowing. This knowing is not the knowing of a subject about a world but the knowing of a subject about itself (or of the world about itself). In other words, it is a knowing in which the opposition and even the separation of subject and object is overcome. Accordingly, it is a knowing which is reflection and which, since it is a total recovery of the object in the subject and of the subject in the object, must be a total knowing of reality, achieved in the form of a total reflection of reality in itself. But such a reflection can take place only if reality itself is reflection, presence to self, or spirit.

Absolute knowing is therefore a knowing in which human discourse achieves coincidence with the reflection of the absolute spirit in itself; it itself becomes the discourse of the absolute about the absolute. Such a discourse must necessarily take the form of a logic, of a rigorous concatenation of constituting moments. The reflection of the absolute in itself is the self-production of the absolute; and this self-production is discourse because it is intelligible through and through; it is concept. Therefore, this knowing is the connection of the concept with itself; and this connection is necessary because it is the very life of the absolute. Thus, the knowing is logic. And it is in the framework of this logic that nature must be understood. It occurs as a moment in the self-constitution of the absolute, corresponding to the positing of the spirit outside itself. Hegel refers to it as the otherness of the idea. Idea "is the true in and for itself, the absolute unity of concept and objectivity."[14] It is the absolute itself.

> Nature has given itself as Idea under the form of otherness. Since, therefore, the Idea is the negation of itself, since it is exterior to itself, it follows that nature is not only exterior in relation to this Idea (and to this Idea's subjective existence, i.e., spirit) but that this exteriority constitutes the determination whereby it exists as nature.[15]

But since it is only a moment, nature is not congealed exteriority;

rather, it is simultaneously exteriority and an overcoming of exteriority. It is only the medium by which spirit returns to itself from its otherness in order to produce itself as spirit; that is, as the being-for-itself of the Idea, or as liberty, defined as "the absolute negativity of the concept as identity with itself,"[16] or finally, as absolute.

> Nature is in itself a living whole; or, putting it more precisely, its movement through its different levels consists in the fact that the Idea *is posited as that which it is in itself*, or, equivalently, in the fact that leaving behind its immediacy and its exteriority (which is *death*), it returns to itself in order, first of all, to become a *living* thing and, next, to put aside this concrete determination in which it is only life and to produce in itself the existence of spirit; for spirit is the truth and the goal of nature and the true reality of Idea.[17]

Thus, for Hegel, nature is shot through and through by one vast movement leading it to spirit. It itself is this movement of the becoming of spirit. Not only does a finality dwell in it, but it also is itself finalization; it is through and through a straining towards an end which is its reason for existing and which exists outside of it. Since everything is connected in a necessary fashion, there is true contingency nowhere. Contingency appears only when we isolate a particular moment. The unification which Kant sought on the level of reason is here achieved on the level of absolute reflection. The Kantian problem of the correspondence between concept and intuition is no longer posed because the concept is no longer understood as an a priori form (correlated to an intuition); rather, it is understood as requiring fulfillment (correlated to Idea, to the pure presence of self to self which characterizes spirit). Concept is self-initiation of motion; it is time. It gives itself its content as it proceeds to develop its articulations. In other words, there is here no more opposition between a transcendental logic and a knowing of content. Logic becomes ontology; it is itself a knowing of content because it is itself the content of its own knowing. It is itself reflection. This transition from transcendental reflection to absolute reflection allows Hegel to reinsert the knowing of nature into knowing pure and simple and to understand nature's finality from the ontological perspective of the absolute. For Kant, finality is only a requirement made by the mind; it is the mind's unifying power as reflected in its work. For Hegel, finality is a law of being. It is the refraction, on the level of nature, of an absolute requirement which is nothing other than the life of the absolute itself.

We know the difficulties these two visions present. Kant, indeed, did develop a philosophy of constitution which can yield a genuine philosophical cosmology. But he thought out his philosophy in the framework of a theory of subjectivity and consequently pretended to derive the forms of constitution from the forms of knowledge. On the other hand, he was not able fully to justify the a priori. The a priori is in essence independent of experience and therefore untouched by any historicity. But how can there be a grasp of the structure of the a priori which is not based on its historicity? In fact, it is on the basis of the historical working of the understanding in classical mechanics that Kant tries to discover the structure of the a priori. It is not surprising that he then finds this structure at work in Newton's axioms. But his theory is called into question by the subsequent developments of physics. We must either abandon a theory of the a priori or else actually succeed in detaching it from historicity.

Hegel, for his part, developed a philosophy of spirit which includes a philosphy of nature as one of its chapters. But if we conceive nature as simply a moment of spirit, how can we really explain that which belongs to nature in its own right? Speaking generally, can we not say that any philosophical theory which pretends to explain nature by something else, and which interprets its finality as a relation to something else, does not furnish a genuine understanding of nature but only of its relation to what it is not? Certainly, we cannot say that nature is completely constituted by such a relation. Is it not necessary to interpret nature on the basis of nature itself; to understand it in terms of its own structure and to allow its eventual relation to another domain to appear only insofar as it is possible actually to show how this relation is expressed within its structure? In other words, a genuine philosophical cosmology must begin by delving into nature itself—not by explaining it from the outside—and by agreeing to go beyond nature only to the extent that the need for such a going beyond can be discerned within its own boundaries. Now this leads us to a reinterpretation of the Kantian doctrine—a reinterpretation that will perhaps allow us to pose the problem of finality differently from the views we have just been considering.

THE FINALITY OF NATURE IN A TRANSCENDENTAL PERSPECTIVE

The central theme of Kantian philosophy is constitution. And this notion connects with a central theme of Hegelian philosophy: manifestation. 'Manifestation' means "rising up in a field of presence."

'Constitution' means "genesis insofar as it gives form (or insofar as it gives birth to a structure)." But the two terms are interrelated; we can think of manifestation only in terms of constitution, and we can think of constitution only in terms of manifestation. If something becomes present, this means that it assumes a configuration, since the function of presence is precisely to effect the separation of darkness and light and thus to make the outline (the figure) appear—therefore, to produce the structure, to constitute. On the other hand, a genesis of structure can take place only when there is a space of occurrences and organizations already available, as when a field of presence is already at hand. Therefore, it is possible only by and in manifestation. Thus, the two central ideas which we just mentioned join with one another.

These two central ideas suggest two questions: How is it possible for the totality to act on its parts? And, on the other hand, is there any movement in nature which carries it beyond itself? These are the two forms taken by the problem of finality in a philosophy of nature. In order to reply to these questions, we must first define the perspective from which they draw their meaning. It is on the basis of an understanding of the meaning of the questions that answers can be given to them.

The perspective in question is that of the a priori or of the transcendental. As we have seen, natural science already puts into play a precomprehension of nature which has the character of an a priori. Here it is only a question of an a priori in relation to knowledge, an a priori which corresponds to a certain idea of nature as a connected totality of processes subject to certain initial conditions. But this idea itself corresponds to a certain mode of giving which it must necessarily presuppose: the idea must be developed in conformity with the manner in which nature presents itself, and nature can present itself only insofar as it produces itself. Thus, the self-production of nature appears as itself the condition for the constitution of an idea of nature and, therefore, as the a priori of this a priori. Here we encounter an a priori of constitution, not of knowledge. Moreover, it is because it is in accord with the self-production of nature that the idea of nature takes the form of an a priori. The a priori of constitution is the set of conditions which govern the self-production of nature. It is transcendental in the sense that it founds the possibility of an appearance of nature, of nature as the system of phenomena. But how can we reach this transcendental domain, and how can we talk about it? The only giving to which we really have access is, it seems, the giving of phenomena; and the knowing of this giving is the science of phenomena. But in the phenomena we grasp only what is given by the giving—its result, the thing itself, *natura naturata*. Our task is to move from the thing given to the giving itself, to the very act of giving, to the very production of the phenomenon, not as determined by another phenomenon but as a phenomenalization—a positing of the phenomenon as phenomenon. We must move from *natura naturata* to *natura naturans*. How is this possible?

It is hard to see how there could be a solution that does not appeal to one of the following: intuition, reflection, or transcendental speculation. The appeal to intuition amounts to the Bergsonian solution. By intuition we can coincide with duration; and duration is precisely the self-production of nature. It is the creative élan which pervades nature. But do we have this kind of intuition? And even if we can have it, can we conceptualize it, can we turn it into knowing? This would seem to be contradictory. We can experience duration and talk about it, but we cannot really think it. The appeal to reflection is the Hegelian solution; thought is the reconstitution of absolute logic and, therefore, the reflection in us of the logic. By systematic thought, we carry out this reflection. And this reflection is identical with the process in and by which nature is produced and overcome. If nature is only a moment of absolute spirit, then a recollection of absolute spirit in itself makes us see the very movement by which nature is pervaded and which produces the forms through which nature appears to us. But is this kind of absolute reflection possible? And even if it is, does it really give us nature in terms of that which characterizes it in itself? It seems that this reflection determines nature only in its relation to spirit, not in itself. It characterizes nature as the being-other-than-spirit and thus makes it the otherness of spirit, but it does not describe the otherness proper to nature in itself; namely, that by which it is other than spirit.

We are left with transcendental speculation. We cannot think the transcendental as a phenomenon because we cannot have a representation of it; we can only think it as transcendental or as a condition of phenomena. But because we do not have a corresponding intuition, we can give our thought content only by reference to phenomena. But is it not contradictory to want to think phenomenalization in terms of phenomena? Do we not thereby give the condition the same status as the conditioned? And is this not transcendental illusion? But it is not a question of treating the transcendental as a phenomenal field. In order to think it we must refer to that which we intuit; but the point is to grasp the intuition by going beyond it, to use representation to evoke the nonrepresentable. In other words, the point is to use the phenomenon as a model and to regard it as such. Now to grasp the model as model is to go beyond it towards the pure form that it embodies; or, again, to grasp its referential movement in a pure way, as separate from the term of the movement. Suppose the representation of a spatial phenomenon *A* is used as a model of a pure transcendental event *B*. This representation includes a reference to a term given in intuition; namely, the spatial phenomenon *A*. Let us put the term *A* in parentheses in order to think from now on, only the *relation* of reference and to direct ourselves, within this relation, to the term *B* which we cannot represent. Here we have the mainspring of our speculative overcoming.

In order to think of nature as self-production we need a model which we must borrow from our experience; for example, the model of growth. In order to think of nature as process, we need the model of movement. If we take our inspiration from the model of growth, how can we think of our two questions? Nature, as *natura naturans*, is the phenomenalization of phenomena; it is the movement which carries the things that appear into the field in which they appear. But this spatial manner of speaking must be corrected. The field of appearing is not given before the appearing itself. Perhaps the most philosphically profound intuition of field theory is that which has us think of the variety of things as identical with the milieu in which they appear. When it comes to appear, the thing bears with it the field in which it appears. It is produced from the very first as spatio-temporal. On the other hand, the unfolding of the field in which things appear is, at the same time, the production of the things for which it provides the possibility of being produced.

Appearing is a genesis. It proceeds from the mere anticipation of the thing to its full presence. This mere announcement of the thing, its pure potentiality, is precisely the field in which it appears as such, insofar as it bears within it (as a field) the becoming, the blossoming, of the thing. On the other hand, the full presence of the thing consists in the achievement of the act of appearing; it is the passage of appearance-in-the making into that which has appeared. In virtue of its presence, the thing is for itself; it is detached from the foundation on which rested the possibility of its arrival. But the power by which it can thus rest in its own supporting space is nothing other than the very force that makes it appear, the influence in it of nature as productive force, as genesis. And since this genesis is simultaneously the genesis of the thing and of the field in which the thing appears, we can say that the movement which carries it out of its field and towards itself is at the same time the movement which brings it into its field. Therefore, in each thing that appears, the entire field in which it appears is assembled and produced while this field produces the thing. This is why the thing is passage; it appears and disappears; it rises up from its field and falls back into it. Or rather, the field in the thing is rendered present—first by letting the thing appear in its own right and then by taking it back into its own pure potentiality. It is by the field in which it appears that the thing is manifested; but it is by the thing that this field itself (which is the field of presence) is itself present. Since the presentative activity of the field is nothing other than genesis itself (nature as production), we must say that there is present in the thing, at the moment when it reaches the height of its presence, genesis itself (nature itself). The thing is produced from a totality, and it is taken back into this totality. But this totality is not a sum of parts, nor a system, nor an idea. It is primordial process, the absolute genesis which is at work in every phenomenal genesis and for which, on the other hand, phenomenal geneses furnish us a model. The influence of the totality on the particular thing is, therefore, the working of transcendental genesis in that thing.

But the passage from potentiality to presence is not immediate. Genesis is not an instantaneous movement but rather a becoming which, as such, has a structure. The becoming referred to here is transcendental becoming, the a priori condition of all becoming. The task of a philosophy of nature is to bring to light this structure of transcendental genesis, to reconstitute the underpinnings of nature as primordial production. The general model for such an analysis is that of the passage of unity into multiplicity; i.e., a differentiating movement. Differentiation is at first the pure dispersion of unity, the production of the pure possibility of the multiplicity. But it is also (and immediately) a regaining of the influence of unity over the multiplicity thus produced; it is a unifying movement. Now this unification cannot be a pure and simple suppression of the posited multiplicity. It is the influence which the multiplicity, in view of its gathering of itself into a unity, exercises on itself. This affecting of the multiplicity by itself is relation. The unfolding of relation in the milieu of multiplicity gives rise to structure. And structure itself is differentiated into spacelike structure and timelike structure. In the latter we find connection, process, nature as melody. In the former we find conditions of stability, configurations, nature as architecture. But structure is still only an intermediary in production. Differentiation, which is the essence of production, can be attained only in the achievement of unification. It remains abstract and is only a condition of true unification. Of itself, the appearing of structure maintains the duality between the relation and the *relata* which it unifies. There is a genuine unification only if, in this unification, the relation recaptures its *relata*—if it produces them from itself. Structure arises from the influence of genesis on the multiplicity. But this influence goes further: it must act so that from structure itself there arises multiplicity and, reciprocally, multiplicity is produced as structure. In other words, structure must become concrete.

But how can there occur a passage from abstract relation to the fullness of structured unity? To put it briefly, there must be a passage through relation so that original unity can come out from its undifferentiated state and be produced as an organized unity, as an achieved system, as nature. Original unity is production itself, but considered only in its pure potentiality. Accomplished unity is produced production; that is, production which has returned to itself. It is not, indeed, a simple extrinsic product, posited outside of its origin. Rather, it is a product insofar as it produces itself and is, therefore, the synthesis of origin and term, the return of the origin to itself. Achieved unity is not simply *natura naturata*, the system of phenomena, but the synthesis of *natura naturans* and *natura naturata*. However, this synthesis does not occur at a single moment. Structure is unified only to the extent that it is itself unifying. It is reduced to the concrete unity of nature posited as a total system only to the extent that it itself produces a concrete unity from the multiplicity that it unifies, to the extent that it goes beyond itself as abstract configuration and moves towards the concrete configurations which enter, as phenomena, the realm of appearances. These concrete configurations are real things in the sense of stable and differentiated unities, phenomenal stabilizations of the universal becoming of absolute genesis. Thus we discover the thing as the intermediary term in the unifying movement, as the provisional product of the unifying genesis.

But how can a configuration have a concrete unity? What is this thing, that which appears? Either it is just another abstraction, a supplementary level in the organization of structures; or it is in some way the whole itself—for only the whole is fully concrete. Since the thing is still marked by differentiating movement (by the otherness of being different), it cannot be identical with the whole. Therefore, it can be produced as concrete only if there is a guarantee of a mediation in it between difference and identity, between otherness and sameness. In other words, the thing must undertake unifying activity in the very heart of its own differentness. It must gather together all differences but without itself abandoning its differentness. The thing must, therefore, sum up in itself the entire universe, the system of all things. It can do this by means of action.

Each thing acts on the entire future. This action is nothing other than the unifying movement of genesis insofar as this movement is gathered into the thing which has come to birth. The thing's action is its contribution to universal unity, to the unification which is at work in it. This unification is at the same time a unification of the thing and a unification of the whole. It is in the particular thing as such that the whole is unified. And it is by achieving itself, by becoming fully concrete, that the thing bears within it, for the instant it is present, the unifying action of the whole. On the other hand, this means that each thing gathers together in its present all past presents, that the entire universe which has preceded it somehow converges towards it in order to give it the fullness of its determinations and, thereby, its concrete unity. Yet it also means that each thing, in its present, contributes to the determination of all future things, that it is itself like material for the concrete syntheses by which future presents will be produced. Thus, each thing is the instantaneous passage of the universe, a center of convergence and divergence, of gathering and scattering; it is the fleeting emergence of the universal stuff, the concresence of duration, the concrete aspect of genesis.

But the concrete synthesis of the thing is still only a preliminary form of unification. As such, it refers back to an absolute unifying movement which is production in the totality of achievement, production insofar as it returns to its origin. As we have seen. the action of the thing is a prolongation in it of all past actions; this action is also the thing's entry into the universal movement, its contribution to the total unification. Therefore, it has meaning only as a provisional concretization (necessarily provisional because partial) of the origin's movement of return to itself. The true concretization is nature as a total unity, as the system of all things or, more exactly, of all actions. But this system is not a stable configuration. It is a becoming, a universal action, genesis itself, nature itself as the incessant production of itself.

We can now understand the meaning of our two questions and answer them. The first question is about the way the whole acts on the parts. We can reformulate it as follows: how are phenomena (the forms by which nature presents itself) affected by the universal and unifying movement of production? The second question is about the eventual relation of nature to another domain. We can reformulate it as follows: Does nature as absolute genesis require a surpassing of itself?

To the first question we have to give a twofold answer. (1) Each particular figure bears within it the action of the whole in the sense that it is the product of a unifying movement which constitutes it as concrete and which in so doing makes it something like an image of the totality. (2) Each figure is only a center of the passage of the universal action and thus is explained only by its reference to the totality in which it must be immersed. Thus, the thing itself is a whole which gives the reason for all the conditions of its appearing. It is a point of convergence for all partial actions. But the thing is whole only to the extent that it is itself called to enter the totality of becoming as part of a unique movement of genesis. Therefore, it is an end and at the same time it looks to an end. It finalizes insofar as it is itself finalized. Thus finality is the demand for concretion.

And the demand for concretion is the very law of manifestation. There is production (entry into presence) only if determinations reach the point of saturation. But manifestation (production) is nature as universal genesis of itself. The demand for concretion is the internal logic of unification. Thus, finality is the action of nature on itself by which nature is produced in terms of its diverse figures in order to be gathered into the unity of its genesis. In this action, each figure recalls all the others, but at the same time each is an expression of the whole, and the whole itself is no longer a figure but the incessant unfolding of the origin, the incessant ascent into the light of day of that which is shrouded in the obscurity of birth.

This gives us the answer to our second question. If nature can and must be understood as genesis or as production, it contains its own meaning. In any case, we can never find within it the necessity of going beyond it. The movement of totalization which is the culmination of genesis is a movement towards closure. The production of nature as fully concrete action gives it the totality of the determinations it needs. It is folded into itself in the way expressed by the majestic image of the eternal return.

However, this is not the ultimate interpetation that can be given to nature on the speculative plane. Moreover, there still remains an absolutely fundamental question: if nature is genesis, does it suffice, in order to genuinely understand the possibility and significance of this genesis, to let ourselves be guided by the model that we have proposed? This model was borrowed from the world of phenomena and therefore from nature itself. It is not amazing that it has led us to think of nature as closed on itself. But this model lets us understand only the "how" of genesis, its manner of unfolding. Must we not go further and try to understand the reason for genesis itself, the fact of genesis, or genesis no longer as a process but as an event?

FREEDOM AS THE FOUNDATION OF THE POSSIBILITY OF NATURE AS GENESIS

It does not seem possible to approach this event by beginning with genesis itself, for when we grasp genesis as process we remain enclosed in it. The event signals a discontinuity. To understand it we must grasp not its result but its pure possibility. Therefore, we must understand this result insofar as it is other than this possibility and, reciprocally, this possibility insofar as it is other than this result. This means that we must move away from what is produced and from the productive act itself in order to grasp the event on which this act depends. Now in this connection we have an excellent guide provided for us by a problematic closely connected to speculation about nature—the problematic of freedom. We can regard freedom in itself as the possibility of the action of spirit on itself, the possibility of ethical action. But this action is expressed; it is translated into gestures, words, and deeds. And, through this expression, it enters the current of the world. A human being is partly spiritual interiority (insofar as he is an acting subject) and partly an element of nature (insofar as he is a body). But the body is not simply a natural agent; it is rather the site of an encounter between the action of nature and the action of spirit. It is the domain of contact in which natural processes are transformed into spiritual determinations and in which, conversely, spiritual initiatives are converted into natural processes. How is this encounter possible? How can spirit be affected by the body and how can it act through the body? Perhaps there is really only one problem here, since spirit can be affected only in an active way, by acceptance or consent. As a result, the real question concerns the action of spirit, regardless of whether this action takes the form of consent or of initiative.

We know how Kant tried to answer the question. He began by showing the possiblity of free action by distinguishing the order of noumenal causality from the order of phenomenal causality. Thus, action has two faces, being at the same time noumenal and phenomenal. As noumenal, it is the absolute beginning of a series, unconditioned by anything outside of itself. It is pure self-determination. As phenomenal, it is a series of phenomena linked together in conformity with the laws of nature. Next Kant turns to the analysis of action as it actually occurs, and when he asks how the encounter between the two orders takes place, he finds an answer in finality. Freedom has a noumenal end. Since it is pure autonomy, this end cannot be extrinsic to it. Freedom's end can be only its own full realization in the form of a kingdom of ends. But to say that freedom is realized is to say that it becomes real, it is actually inserted into the world of things. Thus we can say that the realization of freedom is its phenomenalization. This is possible only if the phenomenal order contains the possibility of being taken up into the order of freedom. Now this possibility is contained in the finality of nature and, more particularly, in the finality of human history, which, before being man's spiritual history, is his natural history—the entangling of human life in purely phenomenal contingency, in the play of interest and passion. The action of freedom can progressively harmonize with and even take upon itself this flow of things in order to make it the material of the realm of spirit, for the finality of the world somehow prepares and (within its order) heralds the finality of freedom.

> We can regard the history of the human race in its general features as the realization of nature's concealed plan for producing a political constitution which is internally perfect as well as—in order to attain this end—externally perfect. Only if history is this sort of realization can nature completely develop all the potentialities that she has placed in mankind.[18]

Now Kant represents the finality of freedom in terms of this political model as the establishment of a universal legislation which includes all men in a unified system or, in short, as a city of ends. Thus, freedom can act in nature because freedom is nature's final end. Popper writes:

> Now assuming that things in the world are beings that are dependent in point of their real existence and, as such, stand in need of a supreme cause acting according to ends, then man is the final end of creation. For without man, the chain of mutually subordinated ends would have no ultimate point of attachment. Only in man, and only in him as the individual being to whom the moral law applies, do we find unconditional legislation in respect of ends. This legislation, therefore, is what alone qualifies him to be a final end to which entire nature is teleologically subordinated.[19]

But in addition to the fact that this inclusion is not drawn from the exigencies of nature but rather from those of reason (and thus, in the end, of freedom), Kant gives us no analysis of the mechanism by which freedom acts on the phenomenon.

Of course we can, as Popper has suggested, "regard the relation between freedom and nature as a relation of control in which the connection is not rigid but plastic; that is, involves at the same time action and reaction."[20] From this point of view, Popper explains, the relation between the organism and consciousness is of exactly the same kind as the relation between consciousness and its "exsomatic" products like language and scientific theories. Now language plays a regulating role by means of its function of argumentation, which Popper distinguishes from the function of enunciation–while the latter consists of forming propositions, which can be either true or false, the former strives to separate true propositions from false propositions. This model suggests bringing into play the notion of information, and we can try to close in on our question by employing this notion.

We could say that the action of freedom, as an action in and on the world, is the conversion of spiritual effort into information. But of what does this spiritual effort consist and how is this conversion possible? When we decide, we represent to ourselves a goal and the means to attain it. These representations correspond to information present in the organism. When the means are actually put into operation, the program of action becomes a phenomenal action (a putting into motion of the body, a concatenation of motions). The spatiotemporal configuration of the motions also corresponds to information. Because of the decision, there has been the transfer of information. That which before was only a program for an operation has become an actual operation. This transfer involves a transformation and therefore a process. But this process is determined by neither the prior nor the subsequent state. These two states remain partially undetermined; the necessity of their connection does not lie in them. This connection was effected by the decision, which of course conformed to the representation of the goal, but without this representation being sufficient to establish the connection. Therefore, there is a gap between the preliminary representations and the execution. In filling this gap, the decision establishes a configuration which is more complex than those which existed before. Thus, it introduces new information into the sequence of the states of the organism.

Of course, the intervention of the decision presupposes that the organism is actually capable of receiving and integrating this new information; that, on the one hand, the organism by itself already supplies a configuration complex enough to be able to support the new configuration and, on the other hand, that it includes enough indetermination to allow the possibility of the new organization. But these conditions are in complete conformity with the general model of the architectonic of nature. It is simply that here, in order to introduce new information, we must appeal not to initial conditions but to something outside of nature. Therefore, we can say that the action of freedom on nature consists in establishing supplementary connections by which it makes appear configurations which were not precontained in the initial conditions of the organism. In short, the action of freedom is an action of synthesis.

But how is this synthesis possible? A synthesis is a higher organization and thus a kind of information. Therefore, we must say that freedom creates information. But how can we represent such a process? Perhaps we can say that freedom by itself is the possibility of articulation. All information is differentiation. It presupposes a break in homogeneity, disassociation. Now freedom is precisely, when it is exercised, simultaneously separation from itself and connection with itself. We could say that it is reflection in itself. Reflection makes difference appear in its pure form and directly connects the disassociated terms in the synthesis which it is. Thus, the internal structure of freedom, to be sure, provides the condition needed for information to be able to appear. But we still must understand how a purely interior act, a purely spiritual connection of self with self, can produce its image outside of itself in the form of information or physical configuration. This problem is more profound and more difficult than that of the relation between meaning and expression. Here, perhaps, we reach a kind of absolute limit of understanding. And we have hardly advanced any further than Descartes when we encounter this problem of grasping interaction at the very instant of the passage from interiority to exteriority (and vice versa).

However, this discussion of the problem is not without value for our subject. This is because free action provides us with a model which we may be able to employ in order to think of genesis as an event. Indeed, by free action we act on nature, but we also act on ourselves. We make novel configurations appear in the phenomenal universe, but we also introduce an increase (or a decrease) in the realm of values. If freedom is a creator of information, it is also a creator of ethical meanings; and, further, it is only insofar as it is ethical creation that it produces information. The information created by freedom does not introduce new laws into nature but rather a meaning which nature by itself and in virtue of its laws does not have. The order of connections is rigid, but the architectonic order is not. This is why a surplus meaning, separate from the universal law of transcendental genesis, can enter into the order of phenomena. If this is so, then it is only by reference to freedom that we can decipher this meaning—not by reference to nature. This meaning appears in nature because it is phenomenalized in the form of new configurations; but as meaning, it is foreign to nature because it comes from the outside. Freedom, therefore, provides us with a model for an action on nature which has a meaning that does not come from nature itself but rather refers to something other than nature. This model lets us think of nature in its otherness and to think of it in its relation to that which it is other than.

Can we not conceive genesis as an event in the form of an act of freedom which would not be an introduction of new information into an already constituted nature but rather an absolute and original positing of the information by which nature is constituted—a positing of the very genesis which, for speculative thought, is nature? This originating act would appear to us as the foundation of the a priori of nature and thus as the ultimate foundation of the possibility of understanding it. But by thinking nature as the product of an absolute positing act (of freedom, of pure synthesis), we prepare ourselves to understand its meaning other than by reference to itself. The evocation of genesis as event breaks the circle of the eternal return. The infinite movement of genesis is the eternal absence of any event. The fulguration of the act renders genesis finite by producing its surging forth as an absolute event.

But if at the very source of genesis, there is freedom as a positing act, then the meaning of our question about the finality of nature is modified. The question of finality turns into that of meaning. It is no longer about the action of the totality on its parts nor even about the possibility of nature's going beyond itself. Rather it is about the arrival in nature of a meaning which nature knows nothing about and the secret of which is possessed only by the positing act. Thus, the true finality of nature perhaps lies well outside of it. For human freedom, as Kant saw, this finality serves as the basis of the initiation of a universal realm of spirits. For the positing freedom, we can think of it as the basis of the initiation of a pure realm of freedom-as-giving. But perhaps the distance between ethical freedom and freedom-as-giving is infinitely more infinite than the distance between ethical freedom and universal genesis. Ethical freedom is merely an anticipation of itself, an effort to coincide with the movement which supports it. Freedom-as-giving is this movement itself, not as tension but as superabundance. It is the pure generosity of the absolute source. In it there is heralded a mystery which belongs neither to nature nor to spirit and to which access is provided only by a revealing word.

NOTES

1. Adolf Grünbaum, "Temporally-asymmetric Principles, Parity Between Explanation and Prediction, and Mechanism Versus Teleology," *Philosophy of Science,* 29 (1962), 146–170; and "Das Zeitproblem," *Archiv Fur Philosophie,* 7 (1957), 165–208.

2. Immanuel Kant, *The Critique of Pure Reason,* trans. J. M. D. Meiklejohn, *Great Books of the Western World,* 42 (Chicago: Encyclopaedia Britannica, 1952), p. 76.

3. *Ibid.*

4. David Hume, *A Treatise of Human Nature, Everyman's Library,* 1 (London: J. M. Dent & Sons, 1941), p. 107.

5. Robert Bruce Lindsay and Henry Margenau, *Foundations of Physics* (New York: Dover, 1957), p. 522.

6. In the statistical theory of Darwin and Fowler, entropy is defined as follows: $S=k \log C$, where k is Boltzmann's constant and C is the sum of the probabilities of the macroscopic states of a given system which are compatible with a given total energy. The probability of a macrocsopic state is equal to the sum of the probabilities of the corresponding microscopic states. Since, in classical statistical mechanics, all microscopic states are regarded as having the same probability, the probability of a macroscopic state is equal to the product of the number of microscopic states which yield it by the common probability of these states. The probability of a microscopic state is itself given by the product of the probabilities of the energy states of the elementary constituents of the complex system being considered; and the probability of an energy state of an elementary constituent is given by the state's degree of degeneracy. This degree of degeneracy is given by the number of elementary states that have this energy.

7. Jules Lachelier, *Du Fondement de L'Induction, Suivi de Psychologie et Metaphysique et de Notes Sur le Pari de Pascal,* 4ème éd. (Paris: Alcan, 1902), p. 70.

8. Immanuel Kant, *The Critique of Judgement,* trans. James Creed Meredith, *Great Books of the Western World,* 42 (Chicago: Encyclopaedia Britannica, 1952), p. 76.

9. *Ibid.*

10. *Ibid.*, p. 469.

11. *Ibid.*, p. 467.

12. *Ibid.*

13. Kant, *Critique of Pure Reason,* p. 205.

14. Georg Wilhelm Friedrich Hegel, *Enzyklopädie der Philosophischen Wissenschaften Im Grundrisse* (1830), Neu herausgegeben von Friedhelm Nicolin und Otto Pöggeler, Sechste Auflage, (Philosophische Bibliothek, Band 33), Hamburg, Verlag von Felix Meiner, 1959, § 213, p. 182.

15. *Ibid.*, § 247, p. 200.

16. *Ibid.*, § 382, p. 313.

17. *Ibid.*, § 251, p. 204.

18. Immanuel Kant, *Kleinere Schriften zur Geschichts-Philosophie, Ethik und Politik,* Herausgegeben, von Karl Vorlander, (Philosophische Bibliothek, Band 47), Hamburg, Verlag von Felix Meiner, 1959, p. 16. ("Idee zu einer allgemeinen Geschichte in weltburgerlicher Absicht," Achter Satz.)

19. Karl Popper, "Of Clouds and Clocks: An Approach to the Problem of Rationality and the Freedom of Man" (The Arthur Holly Compton Memorial Lecture, presented at Washington Univeristy, St. Louis, Mo., April 21, 1965.)
20. *Ibid.*

Time and Finality According to Lecomte du Noüy

François Meyer

Biologists who are partisans of finality generally believe that they have found in Lecomte du Noüy a determined champion of their point of view insofar as biology is concerned. They do not hesitate to develop, in homage to him, the traditional arguments in favor of the finality of adaptations, whether it be the adaptation of the living being to its surroundings, or the complex adaptation of the organs and the functions—a coadaptation which is responsible for the functional unity of the living being. These themes are well known and we will not try to contest their validity. It is questionable, however, that one can proceed to make of them a commentary on the biological work of Lecomte du Noüy and to believe that one thus pays homage to him who would certainly have been considerably surprised by it.

In all truth, Lecomte du Noüy does not seem to be greatly preoccupied with the problem of the finality of adaptations. In respect to adaptive mechanisms he expressly declares that "it is not certain that these mechanisms submit to a finality as was thought up till now."[1] Indeed, adaptation represents for him a state of equilibrium between the organism and the surrounding medium, or a state of equilibrium between the different organs and the diverse functions of the organisms.

The mechanisms of adaptation maintain this equilibrium or reestablish it when it is displaced. For a physicist like Lecomte du Noüy a phenomenon of equilibrium or reequilibration is not a particularly difficult problem. Ultimately a generalization of the Le Chatelier principle should suffice to provide an explanation. We must, therefore, accept it. In the traditional discussions of the eventual finality of adaptations Lecomte du Noüy is not on the side of the finalists and the latter are ill advised to make him their ally.

On the other hand, but from quite another point of view, it is true that the biological philosophy of Lecomte du Noüy is finalistic in its inspiration. But to avoid all confusion, he takes here what seems to him a necessary precaution: his vision of things constitutes "a concept slightly different from finalism."[2] To emphasize this difference he designates his concept by the neologism of "telefinalism."

What does this mean? In all truth, if we wish to determine the views of the author we are considering, we must elucidate the distinction he constantly makes in his works between adaptation and evolution. Evolution is the great oriented derivation of living forms on the scale of the magnitude of geological time. It is a "global phenomenon, irreversible and progressive."[3] Adaptation is equilibrium, is *the* "conservative stability," which is the opposite of evolution, which expresses movement and appeals to the principle of "creative instability."[4] This means that "evolution is essentially different from adaptation," and that "the word and the concept of evolution confirm something more than what is contained in the mechanisms of adaptation."[5] Lecomte du Noüy vigorously criticizes everywhere the "confusion between adaptation and evolution."[6] He writes, "Whereas adaptation blindly tries to attain an equilibrium which will bring about its end, evolution can only elaborate itself through unstable systems or organisms."[7] So that whereas the mechanisms of adaptation are reduced to the general logic of systems in equilibrium—a logic which no doubt derives from finality, contrary to "what was believed up till now"—the global irreversible and progressive phenomenon of evolution calls for a completely different type of intelligibility. It is a temporal oriented vector, a movement transcending the logic of statistics. It implies a principle of transfinality and supradetermination which is responsible at the same time for the movement itself and for its long term orientation, it is a telefinality. It is, therefore, a question of finality, though we can see that it is not for an instant a question of an adaptive finality. On the contrary it is truly an evolutive finality, which is of a quite different order.

More precisely, the question of finality has a meaning for Lecomte du Noüy only in a temporal dimension and as the *problematic of a long term orientation.* It is the problem of the "arrow" of time which alone suggests an appeal to finality. The concepts of time and finality are thus closely linked. The immobile, the equilibrated, the stable, are subject to pure mechanisms and do not suppose any organizing finality. There is no finality in the instant, no finality responsible for organizations as such, *hic et nunc*. Finality appears only with the oriented succession of phenomena. It belongs to the problem of time and acquires meaning only through a review of that problem.

The disjunction between adaptation and evolution which is so strongly marked is also seen by Lecomte du Noüy in a very significant epistemological perspective which is perhaps the most evidently original of his reflections on evolution. First of all, he indicates that we must make a distinction between the mechanisms of evolution and the law of evolution. The mechanisms, on which are based the discussions between Darwinists and their adversaries, do not in themselves contain the explanation of the movement of evolution: "Each one contributes materially and statistically to evolution but without the laws which they obey being similar to those of evolution. The laws of evolution are *teleological* whereas those of the transformation of each species *simply tend toward a state of equilibrium with the surrounding medium*."[8] The tactics of adaptation do not explain the strategy of evolution. Here we come back, in a certain fashion, to the distinction between adaptation and evolution, but the reason for this distinction appears at another level of analysis, as we shall see.

Taking into consideration the "mechanisms" of evolution (which to be sure are mechanisms of adaptation) places research at a certain scale of temporal magnitude, which is infinitely small in comparison with the scale of magnitude of this "global, irreversible, and progressive" phenomenon which is evolution itself. An investigation focused on these mechanisms places in evidence at this scale of magnitude, the laws which rule these phenomena; but linked as it is to its own scale of magnitude, it is necessarily incapable of grasping the global phenomenon of evolution. This can appear only as a result of a transfer of observation to a scale of magnitude having a dimension different from that of the mechanisms. To employ a language now traditional among physicists, there is a disjunction between two levels of observation: a microscopic level and a macroscopic level. The study of the mechanisms is situated at a microscopic level in relation to the evolutive macrophenomenon. It is not astonishing, therefore, that the situation (analogous to the one encountered by the physicist), is the following: not only are the two levels of observation disjointed but so are the types of intelligibility which command the logic suitable for each level, so that *the laws* of the mechanisms do not account for *the law* of evolution. Lecomte du Noüy finds in the classical analysis of Charles-Eugene Guye the epistemological principle which confirms him in this vision of things: it is the scale of magnitude of the observation which creates the phenomenon.[9] We, therefore, have to deal with a disjointed phenomenology. On the one hand the observation, and on the other hand the putting in evidence of the laws, cannot be effected at the same time on the two levels. So as to give to this principle all its authority, we can compare it to the complementary principle of Bohr, according to which the very observation of phenomena tributary to a double line of sight prevents focusing at the same time on two complementary and irreconcilable aspects. Or to use the language of the theory of information, everything takes place as if all information acquired at a certain level decreased to the same extent the possibility of information becoming unified at the level of generalization. It is, therefore, understandable that the most careful elaboration of the law which is proper to the level of micromechanisms is incapable of revealing the evolutive macrophenomenon and the law which governs it.

The originality of Lecomte du Noüy consists in having applied the general principle of epistemological disjunction to the distinction between the levels and the scales of *temporal* magnitude of observation. "The imagination of man is a prisoner of the unity of time which he perceives directly: the mean duration of a human life. . . . He knows, however, that the unity of the time of evolution is immensely greater, but he is not yet developed enough to adapt himself to this cosmic scale."[10] Therefore, it is in fact by freeing oneself from the limits imposed by the temporal scale of magnitude of the observation of the micromechanisms that one becomes able to bring to light the macroscopic, "global, irreversible and progressive" phenomenon of evolution. On this new scale the phenomenon of evolution, as an oriented derivation of living forms, becomes evident and transcends all considerations of the phenomena put in evidence (in particular by neo-Darwinism) on the scale of magnitude of the transforming mechanisms. "Evolution itself thus appears a realm in which adaptation no longer has any meaning."[11] In simpler and more strictly methodological language let us say that the unit of time to which the biologist usually gives his attention is that of experimental research, whose procedures are necessarily—and legitimately—limited in time. At this scale of magnitude it is evidently impossible to perceive phenomena whose scale of magnitude attains 10^6, 10^7, 10^8 years and more, which is the scale of magnitude of geological times and the time of evolution. We therefore understand why a great number of biologists still remain opposed to the idea of a global direction of evolution, and therewith to any allusion to a "telefinalist" hypothesis. It is doubtless the paleontologist, whose investigations put him directly in contact with the scale of magnitude of geological and cosmic times, who is by contrast more open to the idea of a progressive evolution.

These considerations which are fundamental in Lecomte du

Noüy's thought find a positive confirmation in the following reflections of the geologist, M. Gignoux:

> Geologists now unanimously admit that a million years is the unit that must be adopted to see the unfolding of purely orogenic phenomena, the progression of drifts, the flexures, the deformations of hard rocks. . . . We sought, unsuccessfully, to explain the flexures and the drifts by basing ourselves on experiments on a scale of time which is necessarily that of ordinary life. . . . Now the physical and particularly the dynamic properties of a material, those that govern the mechanisms of its deformations, are essentially variable according to the scale of time with which this same material has been considered. On the scale of a million years the hardest rocks resemble fluid plastic matter and can flow even when submitted to only a feeble pressure.[12]

The author does not hesitate to compare this situation to that which is specifically that of evolutionary problematic. Geneticists try to put in evidence the mechanisms of evolution but "they study living beings with the methods and in the framework of the time of *Homo faber*. On that scale of time it is probable that the phenomena which would reveal the key to this mechanism are too subtle to be perceived or put in evidence." One finds it significant that a science which is completely positive and not affected by the prejudgments which so seriously obstruct the study of the problems of life, of itself accomplishes the epistemological mutation which leads it to distinguish, in its own problematic, two levels of phenomenal legality that are irreducible. In this light, the views of Lecomte du Noüy cannot fail to appear as being very positive and capable of casting light on the general problem of evolution. We can no longer be astonished at seeing the macrophenomenon of evolution obey "another law" different from the laws which govern the evolutive mechanisms on the microtemporal scale of magnitude as established by classical research.

The same idea could be presented under another form, which is also instructive. When Lecomte du Noüy affirms that "the laws which the mechanisms of evolution obey are not really identical with those of evolution which dominate and correlate them,"[13] he gives to the law of evolution a dominant and eminent character with respect to the multitude of mechanisms which constitute the proper matter of the evolutionary process. The idea thus emerges that the pure accumulation of events of the evolutive mechanisms and situations take on a macrotemporal "form" which embodies them and seems to give them directional unity. Everything occurs as if the evolutive law draws into its own movement, like a kind of suction of air, the whirlwind of the evolutive microphenomena. In a more positive manner Lecomte du Noüy proposed to give the name of *enveloppe courbe* to the irreversible, progressive global form taken by the infinitely complex mass of evolutive events. Now, it actually seems possible to put in evidence the existence of such enveloping *courbes* by positive methods and in a manner infinitely more concrete than would be supposed possible by reason of the abstract character of these considerations. Thus a great number of significant variables show a positive and regular variation in the course of evolution which can easily be put in evidence by placing them on the axis of paleontological time. The increase of the number of species (A. Cailleux), the positive variation of the encephalitic indices of Dubois or Anthony, a great number of biochemical variables that witness to a biochemical evolution (Florkin), are inscribed as a function of the time of evolution, following the information given by paleontology on perfectly identifiable, regular *courbes* the form of which is that of a typical acceleration.

It therefore seems that the fundamental intuitions of Lecomte du Noüy, concerning a logic proper to evolution as a global phenomenon and integrating in an *enveloppe courbe* the complex events of the history of life, have a positive consistency which can appropriately support and promote a systematic investigation conducted with a rigorous scientific methodology.

The originality and fertility of Lecomte du Noüy's concepts of evolution derive essentially from the basic intuition that the temporal aspect of a phenomenon is susceptible of a scientific and objective investigation. He thought in terms of duration and gave the time variable an epistemological consistence which is often denied it. It is only appropriate to recognize that theoretical physics has carried the concept of time very far; but it is paradoxical to ascertain that in biology the situation is not nearly so favorable. Though the biologist repeats that every biological phenomenon finds in its temporal dimension its very consistence and that every biological reality is necessarily rooted in becoming, he seems carefully to avoid in his research the explicit elucidation, for its own sake, of the temporal rate which is distinctive for the phenomenon he is studying. A proof of this, for example, is the research in the phenomena of growth: if ever there were phenomena for which the time variable appears to be fundamental and constitutive it is certainly this. Yet what do we see? The research of the specialists deals essentially with the *factors* of growth, with its *mechanisms*, or with putting in evidence allometric coefficients linking the relative growth of the organs *as functions of one with another*. As to the variations themselves, they are generally ignored *as a function of time*. Everything proceeds as if the biologist, after having agreed to the principle that every biological phenomenon unfolds in time, carefully turns away from any investigation which takes time into consideration. The temporal aspect as such no doubt seems to him to be a kind of epiphenomenon, or perhaps a kind of more or less Bergsonian contingency. On the contrary, the fundamental idea of Lecomte du Noüy led him to focus the investigation, both in experimental research and in theoretical thought, on the temporality itself of the biological phenomena. In this respect his experiments, now classic, on the cicatrization of wounds mark in their experimental rigor the first and most significant applications of this fundamental intuition. One can see here in germinal form the ideas which would be developed later in the realm of evolution.

We all know under what circumstances these experiments were conducted. The aim was to establish a rigorous method for testing the relative efficaciousness of treatments applied in military hospitals to cicatrization. The experiments in this realm deal essentially according to a causal logic with the determination of the mechanisms of cicatrization. In this perspective, cicatrization is only the effect of a certain number of factors. Therapeutics can intervene to modify these factors, or to introduce new ones, with the aim of obtaining cicatrization more surely or of accelerating its course. But no attention is paid to the temporal speed of cicatrization as such or to the *enveloppe courbe* of the multiple mechanisms which insure cicatrization. Lecomte du Noüy, disregarding these mechanisms, applied himself to putting in evidence the temporal form of the phenomena. He devoted himself to a rigorous formulation of it as a function of time taken as an independent variable, and thus determining the evolution for a spontaneous cicatrization of the surface cicatrized taken as a dependent variable. Well-conducted observations very clearly lead to an evolutive equation which is perfectly identifiable and can be expressed in a simple mathematical formula. The phenomenon of cicatrization, putting aside everything we know and everything we ignore about its mechanisms, finds its temporal identity and this defines itself as surely as do any other of its properties.

The most interesting fact about these conclusions is well known. When the chemical study of the respective efficaciousness of diverse therapeutical interventions was left to experimental inquiry it came to a standstill. The spontaneous equation of evolution plays the part of an objective reference, enabling the evaluation of the acceleration or retarding effect on the different interventions of a causal type.

It is then evident that the knowledge of the true phenomenology of the phenomenon studied constitutes not only a significant theoretical elucidation but, what is more, the preliminary condition for every inquiry of a causal or mechanistic kind. The intention to transfer research to the sole temporal aspect of the phenomenon and to ignore by a methodological decision every consideration of the causes of the phenomenon, is shown to pay off and to constitute a moment of objective elucidation. The highly significant fact of the time variable is thus put in evidence as well as the significant existence of an *enveloppe courbe* identifiable in itself. It can be seen that this situation proceeds from the same principles that are later found in the concept of evolution. The positive character of this experimental research constitutes the very foundation of the more theoretical-analytical steps taken concerning the problem of evolution.

Taking a closer view, it is in still another way that temporality manifests its significance. Indeed, the equation of the evolution of cicatrization contains a coefficient *k* which varies with the age of the organism in question. Time, therefore, intervenes on two temporal scales: first, on the scale of the phenomenon of cicatrization itself, then on the scale of the evolution of the organism. We, therefore, ascertain a double intervention of the time variable, a distinction of two scales of time, and finally a conjunction of these two scales in every concrete phenomenon of cicatrization.

From this we can conclude that the impressive originality of Lecomte du Noüy resides in his purely temporalist approach to biological phenomena. He proved that such an approach not only conforms to the essence of the biological phenomenon, as duration and variation, but is also capable of giving rise to experimental and positive steps which witness to its objective validity. On the other hand, we remember that it is again the dimension of time and the presence in it of an oriented arrow, which invokes the problematic of finality and leads to the telefinalist hypothesis. Whether it is a question of a concrete investigation, methodologically faultless, or of a theoretical concept of the conditions of intelligibility of evolution, it is always the dimension of time which is predominately in the thought of Lecomte du Noüy and which gives it its particular form.

We cannot, however, limit the thought of Lecomte du Noüy to considerations taken from the positive sciences. His books are often and above all read as a spiritual testimony, which is at the same time humanistic and theological. If he admits that evolution is a "global phenomenon, irreversible and progressive," it is perhaps in the hope of putting into perspective the final product, the most complete, the most "improbable"—man in his full stature. The secret of *L'Homme devant la Science* is surely *La Dignité Humaine*. Furthermore, the reflections on evolution, if thought of in terms of their bearing on the concept of hominization, come to no other conclusion, once the explanatory pretentions of pure mechanism are established. The only hypothesis which Lecomte du Noüy considers compatible with the order of magnitude of the phenomenon is the "telefinalist hypothesis," which necessarily leads to the "hypothesis of God." Lecomte du Noüy formulated this conclusion in the secret verity of his conscience, and the question of knowing whether it was in virtue of the scientific rigor of his thought or, on the contrary, of his desire to transcend certain exigencies of objectivity necessarily remains ambiguous, as does, since all time, the relations between reason and faith. But for many who study his works, he will always remain the witness of a humanity which is forever situated between the problematic of its knowledge and the metaproblematic of the absolute—which is of another order.

NOTES

1. Lecomte du Noüy, *Human Destiny* (New York: Longmans, Green and Co., 1947), p. 86.
2. *Ibid.*
3. *Ibid.*, p. 78.
4. *Ibid.*, p. 89.
5. Lecomte du Noüy, *La Dignité Humaine*, (Paris: La Colombe, 1952), p. 71.
6. *Ibid.*, p. 56.
7. *Human Destiny*, p. 90.
8. *Ibid.*, p. 85.
9. *Ibid.*, p. 11.
10. *La Dignité Humaine*, p. 24.
11. *Ibid.*, p. 96.
12. M. Gignoux, *Strategraphic Geology* (San Francisco: W. H. Freeman, 1955), p. 652.
13. *Human Destiny*, p. 85.

Some Observations on the Philosophy of Science in France

Edouard Morot-Sir

Cultural exchanges take place on all sorts of levels of the intellectual, artistic, and moral life of peoples, and currently respond to a general need for communication within the framework of a world civilization. A meeting such as this takes place at the fundamental level—by this I mean the level which should serve as a frame of reference for every cultural experience between nations. In effect, our present effort at reflection, comprehension, and communication affects this frontier zone which at this time in human history plays an essential role in the development of cultures: it is epistemology. In its traditional and actual sense of the word this is the "theory of science."

I have thought, then, that the best way to pay tribute to Lecomte du Noüy, a great scientist and a profound philosopher, would be to discuss the present state of the theory of science. Permit me these few observations inspired by my dual experience as a student of philosophy who, despite the existentialist trend of the forties and fifties, has always felt that the problem of science was central to every philosophical reflection, and then as a diplomat concerned with cultural relations between the United States and France who has more than once tried to analyze and understand the values which draw us together and those which separate us as peoples. I am more and more convinced that a loyal comprehension of differences necessarily leads to esteem and friendship, sometimes more surely than the superficial affirmation of mutual trends.

In the nineteenth century and in the first part of this century, it has often been felt that science, due to the universality of its method and the necessity of its laws, escaped the fact of cultural plurality or even cultural pluralism. This brings to mind Pasteur's famous and generous saying, "Science has no native land"; and after 1918 a French philosopher hoped to found the Society of Nations on the objectivity of scientific truth. We no longer foster this sort of illusory belief, but I wonder whether the roots of that illusion are not still alive and whether we do not continue to accept a conventional image of science which keeps us from noting its authentic and human traits.

In effect, for more than a century a sort of intellectual solidarity has been established between positivism and science, as if positivism were the only legitimate theory of science. Despite some protestations, this bias is still very much alive today and you will observe that even adversaries of positivism, without fully realizing it, accept the positivist interpretation of science, and particularly the reduction of all epistemology to a double theory of deduction and induction. I give my unreserved approval to Karl Popper in his campaign against what he justly calls inductivism. In this way logicians who, since a generation ago, have become positivist logicians have constructed a schema for science which has no reality other than in their own imaginations, or at least in an ideal of science, the purity of which is respected only by the logician. I do not mean to say that the analysis of the problems of science formulated by the logician is false or artificial, but I am afraid that it is real only for him who works from the results already obtained by the scientist. The positivist logician puts science into a rarefied air which, I believe, would be unbreathable for the scientist in his laboratory and even for the mathematician at the blackboard. To solve his specific problems, which have but little to do with problems of logic, the scientist thinks with all of the resources of his individuality and with all his soul, as Plato said of philosophers. And in order to arrive at a solution amidst the tumult of research, the scientist is capable of making pacts with everyone, even with the Devil! Thus science, like all human thought, furthermore, is the result of a unique marriage between the universal and the individual. Here also, as in art or literature, the work bears the indelible mark of its author.

This epistemological relationship between universal and individual is itself found within another relationship which reveals the total character of the scientific act. In the present state of human civilization, every scientific effort is more or less patterned after a tradition. There is a diversity of traditions leading to science, and each of its traditions is situated within the framework of a particular national mentality. In other words, a scientific result, no matter how objective it may be, has (if it is considered in its totality) a historical and a national dimension. There is a French aspect of science just as there is an American, a German, or a Russian aspect.

And now I should like to discuss some traits of what could be called the French scientific spirit. You know that for generations since the birth of the modern era France has had a sort of cultural hero who became a popular hero and who belongs to the national character: René Descartes. Cartesianism has served as a touchstone and even as a frame of reference. It has directed the course of our system of education, both literary and scientific, and each of us more or less consciously defines himself as being for or against Descartes. This applies to the theory of science also and it has led to a certain difficulty of communication with Anglo-Saxon thought. Let us look beyond the banal Cartesianism which is often summed up as a narrow nationalism with a positivist tendency, and we shall discover an epistemology dominated by mathematics, the other sciences being defined in terms of their power and degree of mathematization. Reality should accordingly be progressively interpreted in terms of mathematical relations and proportions. Descartes dreamed of a universal mathematics which would extend the Platonic dream and which would serve both as the model and the structure for all knowledge. A characteristic which for generations has marked French sciences is that this idea of mathematics is conceived in opposition to logic and consequently to Aristotle. It has been said that reasoning in the form of the syllogism is sterile and has nothing in common with mathematical reasoning which, on the contrary, is productive and a sign of real progress in thinking.

The problem at the heart of the theory of science is thus that of this interpretation of mathematical progress; that is, a new form of reasoning in comparison with the syllogism of the logicians, which belongs to the realm of common thought. Such an epistemological attitude is found, for example, in Henri Poincaré, who refers to *recurrence* to explain the development of the mathematical universe. Thus the hostility of Poincaré and his colleagues toward research in scientific logic can be understood. In our time there is the Bourbaki movement which, in its immense effort to axiomatize and formalize mathematics, retains this mistrust of logic and has always refused to refer mathematics to anything other than itself. One of the last testimonies of this latest Cartesianism is evident in Jean Cavaillés, a philosopher and mathematician who, before 1940, worked with David Hilbert. In his posthumous book on the theory of science he poses the problem of science in simultaneous terms of necessity and progress. His condemnation of Russell and of logical positivism is revealing. These philosopher-logicians are condemned because they are unable to establish and to justify at the same time necessity and progress. Cavaillés, in the last pages of his unfinished book, suggests a solution which is not far removed from the Cartesian dream of a universal mathematics. He sees developing in mathematical thought a sort of autonomous structure. He even uses the word *dialectic* in a sense which is not Hegelian or Marxist but whose purpose is to describe a thought process which is not reduced to logical and analytical form. We can consequently understand the criticism sometimes directed at French science for being too exclusively theoretical and for according only secondary importance to experimentation. But this criticism is unfair; the relation between mathematics and experimentation simply does not have the same meaning in this ideal of science and in the empirico-logical conception of it.

Now I should like to turn to another aspect of this epistemological Cartesianism which has caused a permanent crisis in France since the nineteenth century. We know about the Cartesian dualism of matter and mind and of the consequent refusal to accord animal life a reality distinct from material properties. This is the generalized mechanism, an extension of universal mathematics. French cultural thought has been divided since the advent of biology in the nineteenth century, and especially since Claude Bernard, as to the problem of the determination of the originality of life and that of the interpretation of finality. I do not need to tell you of the major role played by Bergson in the early years of this century, and later by Lecomte du Noüy.

I have a lively recollection of my philosophical studies at the Sorbonne. Within the Cartesian atmosphere which I have just described, our education was dominated by three closely linked problems: (1) How is mathematical progress to be explained? (2) Does finality exist, and if so, does it belong to the realm of scientific or of metaphysical knowledge? (3) Is a science of man possible? These three problems are united in a new understanding of the mathematical universe. Within the French cultural framework, then, it is possible to be both Cartesian and anti-Cartesian (Pascal), and we see that a Cartesian conception of mathematics is in no way dependent upon a positivistic philosophy.

And what of the future? Is France condemned to this cultural system of Cartesian references? Is not present-day science, carried away as it is by its fantastic experimental developments, becoming separated from traditional and national structures? The numerous exchanges between scientists throughout the world could lead us to think that we are witnessing the birth of a universal theory of science. And this is the probable conclusion for a number of problems defined within the limits of a particular field. But as for the general problem of science dominated by that of mathematics, I believe that it will remain associated for a long time to come with a certain cultural plurality which has no value in itself except in these incessant exchanges which are being established between neighboring systems of reference.

Permit me to return once again to this French experience which I know for having lived it out. After 1940 the theory of science was put aside in the name of the theory of existence. It is interesting to note that philosophy in the form of existentialism rejected epistemological reflection, treated science as a secondary existential phenomenon, and isolated itself in a psycho-sociological theory of consciousness. But since 1950 we have witnessed a revival of epistemological research and I am happy to greet here some of the representatives of this new philosophical effort. It is interesting to note that current questions center largely around time and future as time's essential dimension. We should also note the importance of methodological and philosophical research regarding the problem of life. Lecomte du Noüy was one of the first to stress intercommunication between the physical, chemical, and biological domains. The problem of finality is the order of the day, as is consciousness; traditional problems, you will say, and this is true, but treated in an entirely new style. I also believe that the notion of function in its diversified aspects will have new interpretations which will permit a stronger liaison between mathematics, biology and social sciences.

But let us not be too hasty in our predictions. The history of human invention is made of surprises which are then rationalized. The main object of my few observations on the theory of science has been to stress some points which are often neglected by professional epistemologists. First of all, is it not a fact that there is no pure science in the sense of pure intellectual and methodological construction? Every science has cultural roots, and although it retains its objectivity and its universality, it bears the mark of a certain perception of the world which is both that of a group and that of an individual. Second, modern-day science with its power, wealth, and diversity is dominated by its concern with establishing a certain unity from out of the multiplicity of its domains. This, it seems to me, is why the notions of function and consciousness are being given new interpretations, and also why the dynamic Cartesian schema of science remains alive, even if the solutions are anti-Cartesian. Finally, these historic and cultural roots of science are proof that the theory of science cannot be sufficient in itself. The dream of a logical autonomous totality is unfeasible; it would result in the death of man's personality. There is no necessary link between science, materialism, or positivism. On the contrary, science goes hand in hand with man's metaphysical invention. I should even say that there is no great scientific thought without a metaphysical impulse—in other words, without man's total involvement and deep awareness of the universal destiny in which he participates. This is to say that, consciously or not, science implies a religious source.

In brief, neglect of the religious source and of the cultural structure of science risks making of science and its by-products a huge machine which turns round and round indefinitely with neither reason nor goal. This problem of the human position of science, which could appear to be the ultimate question of every epistemology, is not only metaphysical but also political. If it is true that we need a new theory of science, it is urgent that this theory extend to a political system. But who will be responsible for this system? It is difficult to predict. It is certain, in any case, that human government must change and also that before we proceed further, the scientist himself must gain a new insight into science. And it is my hope that a gathering such as this one will serve to pave the way for this new insight. Doing so, we shall remain faithful to the great lessons of Pierre Lecomte du Noüy who by his works and his life has demonstrated the deep unity of science, philosophy, and religion and has shown that the theory and the practice of science is fundamentally an act of faith.

Introduction to a Debate on Philosophical Exigence

Henri Gouhier

In this introduction to a discussion on philosophical exigence I would like to present three points of view. The first will be concerned with the development of philosophy, the second with its birth, the third with the nonphilosophical sources of philosophy.

The first question is: How does the philosophical historian reason? What does he find at first glance when he looks at the history of philosophy? The answer is that at first glance the history of philosophy would seem to be a crushing blow to philosophy. Is it not a panorama of ruins? Is it not a cemetery of systems, and (to continue the metaphor) is not the historian of philosophy the guardian of the cemetery who comes to put the tombs in order and cover them with flowers every year or every hundred years?

If, then, there is a philosophical exigence in the minds of men, seen in the perspective of history, it would run the risk of seeming to be an unreasonable exigence, for it would appear to be pledged to search in vain. Moreover, is an unreasonable exigence still an exigence? Should one not speak therefore of a provisional exigence, coinciding with a kind of adolescence of man and destined to disappear as soon as he will have become conscious of its vanity? I believe that this impression is associated with an incorrect image of the history of philosophy, or more exactly with an image falsified by a schema of progress which, rightly or wrongly, was suggested and formed by the history of the sciences.

I recall here a statement by Alexandre Coilleret: "The history of science is certainly not a dead history: it is *grosso modo* the history of dead things." In his opinion, the Copernican or Newtonian astronomy no longer interests anyone, having no current value of its own. For this reason he thought the history of the sciences differs radically from the history of philosophy. Philosophy holds, rightly or wrongly, that the thought of Aristotle or of Plato has value.

Of course we understand the meaning of Coilleret's sally that the history of the sciences no longer interests anyone—anyone (he naturally implies) except the historians of science or the scientists curious about the past of their science.

Let us go further. Even if we decrease the significance of the Coilleret statement by considering that science recuperates its past or that it preserves something of its past, what I wish to retain is that if we speak of the principle of Archimedes we speak of that principle in an actual context which is no longer the context of what science was in the third century B.C. Hence the deep difference insofar as the history of philosophy is concerned. In philosophy indeed it is truly the Aristotle of his time we consult. It is always the Plato of his time that we also consult. But above all with philosophers one cannot employ *grosso modo* any of the images in which common sense could express scientific progress. Bergson goes no farther than Plato or Descartes. Certainly if Bergson constructs Bergsonism it is because he thinks he is saying things that Plato or the other philosopers had not said. And just as surely, at the core of every philosophy is a rebuilding of the philosophic past. The whole final chapter of *Creative Evolution* retraces the past history of philosophy as seen by Bergson in the light of Bergsonism.

But from the point of view of the historian, or of one who is not the constructor of a philosophy and does not look at the past of philosophy in the light of his own thought, we must recognize that such a line cannot be drawn.

What, then, is the situation which confronts us? I think that in order to come to grips with the problem, we should reflect on the notion of culture. Culture has perhaps been too willfully understood as an essentially pedagogical entity. After all, what do we mean by "cultivate"? Cultivate a plant, for example? Of course, this brings us back to the pedagogical point of view. That is why one does not tire of citing the formulas always cited by people with an excellent memory—such as "Culture is what remains when one has forgotten everything." I think culture is something else. It is first of all the historical dimension of our existence. One must look at it first from the angle of reality: "I think, therefore, I am" said Descartes. But even though I am in the universe of nature, I think in the universe of culture and these two universes are as inseparable as the "I" and the thought in the individual who says "I." The musician composes in the world of musicians, the painter sees in the world of painters, even when he looks at a landscape; the philosopher thinks in the world of philosophers, if only to reject what the others say, if only to put himself inside of society. And this historical dimension of human existence as a musician, as a painter, as a philosopher, is precisely what history reveals.

It is in this perspective that we can probably speak of a certain progress of philosophy and can reject the pessimistic and despairing outlook given at first sight by the history of philosophy. This progress which is specific to philosophy is not in knowledge. It is in reality itself. The progress of philosophy does not consist in such facts as that Bergsonism would go farther or see farther than Kantianism or Platonism. The progress of philosophy, we believe, is that there has been—there is now in our universe, in the universe of culture in which we live—a Bergson, and in the universe of Bergson there is Kant, but in the universe of Kant there could not be Bergson.

What is progress in philosophy? It is the reality to which the subject of philosophy belongs and which continually grows. It is not a chronology which would undertake to determine values, on the basis that that which comes now is necessarily better in its place than that which came before. Though the average young mathematician in a college class knows more things than Archimedes with all his genius could know, one cannot conclude that the greatest contemporary philosopher, because he is contemporary, is superior to Aristotle or to Plato. But the contemporary philosopher—even if he be mediocre—is in an infinitely richer universe than the one in which Aristotle or Plato thought, inasmuch as between Aristotle, Plato, and ourselves there lies so very much of history, the great names in which I need not mention. Seen in this perspective it appears that far from giving the impression of a graveyard, there is on the contrary nothing more exalting than to see, in every period of history over the centuries, a continual and indefinite resurgence of the philosophical exigence. That is, then, the first point of view suggested in approaching the philosophical exigence.

But I think you would reply by saying that this final exaltation is only an amateurish esthetic point of view. You will say: we find this beautiful, we go forward in the history of philosophy as we would in a museum. I gladly accept the comparison for I never have a more intense appreciation of life than when I enter a museum. I feel myself in a world infinitely more alive than when I am in the street or in the subway. You will, no doubt, say to me that this is an esthetic point of view. But then you must explain why the philosophical exigence of which I have spoken has profound reasons for resurging indefinitely and for being, as a result, indefinitely new. To answer this question one should know what philosophy is. Precisely. What is philosophy?

The historian of philosophy finds himself in the paradoxical situation of literally being unable to say of *what* he is writing the history. I think one may hold that there can be different kinds of neutral concepts. For instance, there can be a neutral concept of music. I mean by this a concept that does not involve a musical theory. The frontiers of music may be difficult to define, but even so the historian of music knows very well how and where to encounter the people whose history he is writing. This is not true in philosophy. There are no neutral concepts of philosophy, for every definition of philosophy must depend on a particular philosophy. There is, therefore, no neutral concept which enables the historian to define philosophy and to seek in the light of this definition the people of whom he could say that they are philosophers. For let me repeat once more, every definition of philosophy arises out of a particular philosophy.

In other words, every definition of philosophy is based on some kind of compromise. Take Descartes, for example. What does he do when he approaches philosophy? Consider the third part of the *Discours de la Méthode.* After he has put aside the truths of faith, he arrives at a very precise definition of philosophy which is very limited in scope. There will be philosophy only when reason and experience are in question.

Or consider Malebranche at the same period in history. For him there is only one "source" of truth: the "word." But the word speaks to us in two ways: by reason it teaches us mathematics, the moral truths, and it is the internal "master" that we discover in ourselves. But this same word speaks to us through the Scriptures, this time in such a way that it will be heard and understood by all, even by people who are not philosophers. But then, as there is only one "source" of truth, why do we (who participate in reality by these two paths) not join them together? Why not write a Treatise of Nature *and* of Grace? (Note the importance of *and.*)

This is a radically different concept of philosophy from the preceding one. From this point on I need not insist on the number of problems which can arise. Why include Pascal in a history of philosophy? He did not write a single treatise which deals with what is called philosophy; and when he came in contact with philosophy he discredited it. You can see how many problems arise because we cannot "state" what philosophy is since we do not have, I repeat, that kind of neutral and commonplace concept which could, from the point of view of a consensus, permit us to say just what philosophy is.

I think, therefore, that if we look at the history of philosophy in this perspective we ascertain that no matter what the prefatory declarations may be, the whole history of philosophy has been developed in an empirical fashion. We are in the habit of including Pascal in the history of philosophy. We do not ask whether we should or should not do so. We simply do. In fact, I believe that whether it be Pascal, Descartes, or Spinoza we are considering we can see two intentions in all those who have considered themselves philosophers or whom we call philosophers. I should perhaps say "intentionalities"—it would sound more "learned"!

Let us consider the first intention. I think philosophers are people who look at—who try to see what really is. It seems to me that what underlies all the great philosophies is the paradoxical idea that we do not see what is in front of our eyes, that we do not know how to look at what we see, that what we call real is precisely that which should be discovered, and that after all we will perhaps only arrive at approximations. There is here an intention which aims to attain to the real and to see. That is why I believe that the most neutral expression of this first intention is the phrase "vision of the world."

There is also a second intention which seeks to translate this vision of the world into coherent speech. Thus to translate it means to translate it in such a way that there will emerge a kind of order and that, no matter how fluid, this order will develop a language which can communicate a vision of the world.

I will insist on this second point because one of the errors of the younger contemporary philosophy may be the fact that the problem of communication is overlooked. The philosophers that interest history are people who have spoken and people who in general have written. For whom and for what reason did they write? They certainly did not write for the professors of philosophy. They wrote for mankind. That is why the problem of communication is so important. I will refer again to the *word* (rather than communication, because of its ambiguity). The *word* is at the same time thought and word; and this corresponds to the second intention of philosophy.

All this should of course be more accurately defined. Roughly this desire for order—order which allows a certain communication—is probably what distinguishes the vision of the world of the philosopher, who might be Bergson, from the vision of the world of a poet-philosopher, who might be Paul Valéry.

This double intention is a fundamental notion which seems essential in philosophy and, moreover, accounts for the disagreement between philosophers. When philosophers of the same epoch oppose each other, it is not a question of verbal meaning. What is behind the words used is important; one can always agree on words. But if we consider the profound dissensions between philosophers, we are always brought back to what I pointed out just now; namely, that they do not see the same reality. It is not a question of the same *profound* reality.

That is why for a Platonician—Descartes for example—reality is not the color of these walls, it is not the sensations which the lights overhead give us. True reality is *extension* and the measurable movements which are in that extension. For Descartes, that is what is real. How could he then agree with his Aristotelian adversary, for whom the real is precisely all this qualitative universe which Descartes rejects and replaces with the species of intelligible universe given him by mathematical physics?

Or we may ask: why was there such a dissension between Bergson and certain of his contemporaries? Is it because some of them remained faithful to the idea that the definition of the real is essentially the unchangeable? For Bergson, on the contrary, the real that he saw, the real that he experienced, was defined by change. In other words, the profound study of the disaccord that exists between philosophers would show that there is something fundamental which is their vision of the world. The great disagreements come precisely from the fact that different minds do not see the same reality.

This thesis of the vision of the world leads us to the third point I wish to discuss with you. This will enable us to understand why the history of philosophy is a continual renewal, why there are always new philosophies, and why the very essence of concrete philosophy is always to be new. If philosophy is first of all vision of the world, how can it escape being subjected to a perpetual renewal since it is linked to a world that changes?

We speak too often as if only two terms had been considered. On the one hand the real world, immense certainly but always there, immutable, always identical with itself. On the other hand, there is the mind of the philosopher who looks at this world and tries to see and understand it. Between the two terms, there is a third term called during the Middle Ages *Imago Mundi*, the vision of the world. The mind of the philosopher works on a vision of the world, the one he gives himself of the world at a certain period, and this image of the world is never the same. This brings us back to what I referred to earlier: "the nonphilosophical sources of philosophy."

First of all, a vision of the world is never the same because of science which ceaselessly gives us new "editions" of this world, editions that have been reviewed and corrected in such a way that in certain chapters the text changes very rapidly. What is more, the world also changes by reason of the fact of man's presence in his history, with his impulses which I will call "religious" (I give this word a very broad meaning, the meaning of "spiritual experience" simply to indicate that there are changes that depend on the "powers of sentiment"). On the one hand, to be sure, the sciences renew without ceasing our image of nature, but we must not forget that this notion of nature is itself an idea of culture. On the one hand, then, the sciences ceaselessly retouch our image of nature, but they renew our sense of the natural—I think we should go far in the direction I here indicate. And what was natural at a certain period was perhaps not natural in the preceding period; and the spiritual experience, religious in whatever form, launches and relaunches man on the pursuit of a more internal being which we would call (to use a term which is noncommital) "trans-natural."

This idea seems to me to be confirmed by the history of philosophy. The great philosophies are not born of reflection on anterior philosophies. It is we, the professors of philosophy, who—when we seek to give a conclusion to an outline of a philosophy—always try to reflect on the philosophical systems. But on the other hand, what is at the origin of the philosophy of the great philosopher? There one finds precisely what I referred to: a look at reality which leads towards what we call a philosophy; namely, once again, towards what the philosopher will arrange, express, and communicate. He has the impression that he is seeing reality in a new way, and because it is new it is brimful of wonder.

The great philosophies are all the products of deeply felt wonder. Think of Aristotle, amazed when confronting the fact that a living being bursts out of an egg. There was an egg, now there is a chick. Think of Descartes, amazed by the purely intelligible reality that he discovered to be more real than the emotive qualities which enable us to appreciate the significance of our senses. Think of Bergson truly experiencing each instant of time, as he says in the *Essai sur le possible et le réel.* Each instant is fresh, each instant is new, and no matter how commonplace it is there in its every presence something more than anything man could foresee.

I think that if one tried to absorb the meaning of philosophy this is what he would see: it is linked closely with science *and* with spiritual experience, whether the thinker be Thomas Aquinas in the thirteenth century brought face to face with the physics of Aristotle, or Descartes confronted with new physics of his time. Twice in France, and in opposing directions, a biological model was substituted for a mathematical model. In the eighteenth century, thinkers like Diderot substituted a biological model for a mathematical model for the sake of establishing a kind of vitalistic materialism; and in the nineteenth century, Bergson reflected on evolution in order to gain a spiritualism for which the vital impulse is also spiritual energy.

The spiritual experience, or religion, is the whole history of the great turning point at the beginning of our era. When the Old and the New Testaments really entered into the world of culture, it was no longer possible for philosophers, even by metaphor, even in analogies, to do what Plato could still do, namely to suggest the intervention of numerous gods. Even by analogy we cannot imagine a philosopher speaking of Jupiter or some other god. At this great turning point a religious idea like the idea of Creation enters into the world of philosophers and becomes rational and reasonable. Yet this idea was unthinkable, I believe, with the mental tools of Greek philosophy.

Finally, and above all, historicity becomes a dimension of human existence and ideas such as those of sin or grace will become for the philosopher a kind of summons to reflection. In many cases these two sources mingle, often because there is at the same time a summons of the science of the period to the philosophers and a kind of other summons which comes from the spiritual life that there is a philosophy.

I spoke just now of Malebranche. It is Malebranche who discovered a convergence of a new science—the science of Descartes, which excludes from physics all the occult qualities—with faith; and he then proclaimed that God creates everything!

I also spoke, too, of Pascal. It is Pascal discovering the convergence of a new science, the one which—to cite the formula of Coilleret—makes us pass from a closed world to an open world, who discovered also the convergence of a science which projected an indefinite world, with a theology which adores a hidden God and forces us to question even the words of a psalm. Can the heavens still sing the glory of God in an indefinite world, at least for those who have no faith? It is then all through history that we see these two sources come to renew, and continually to force, philosophy to renew itself.

It seemed appropriate to me to evoke these three themes in the course of our *Colloque Lecomte du Noüy*. A particularly important circumstance for the history of the philosophical exigence is the fact that Lecomte du Noüy was a man who was not a professor of philosophy but in whom the philosophical exigence arose out of work which once more was not the work of a professional philosopher.

What then do the experiments of Lecomte du Noüy suggest to offer us? Let me take a simple example, the only one I am capable of understanding. What do his experiments on cicatrization tell us? If I have understood correctly, they do not give us a new idea of time. They give us rather a new temporal reality. The idea of time which Lecomte du Noüy labored to reduce to its essentials is a chain of thought adequate to this temporal reality. It is the special time of a certain biological reality. We realize here through a very precise particular fact how a part of the real in the latest edition of knowledge makes necessary the refinement of an ancient philosophical concept.

I think that we could find in all the writings of Lecomte du Noüy examples much less simple than the one I have mentioned. His thoughts on the statistical laws for example, which prolong themselves into philosophical views on a new determinism; his ideas on molecular and granular matter brought to his attention; the most recent physics, which prolong themselves into important philosophical views on the continuous and the discontinuous. Finally there are his continually renewed reflections on evolution.

On the other hand, I do not think that it is by chance that Lecomte du Noüy so often cited *L'Avenir de la Science* by Renan. We know that this book, published in 1880, had been written in 1848. It seems to me that almost exactly a century later in the different works of Lecomte du Noüy everything seems as if he had wanted to write a new *L'Avenir de la Science* but this time for the middle of the twentieth century.

What is striking is that, here again, reflections on the man who practices science raise all kinds of data to be retained which concern the man who does not become a practitioner in science. I am thinking notably of the invitation to surpass oneself which Lecomte du Noüy experienced in himself when he heard Kierkegaard speak of the passion for the infinite or when he liked to recopy this line: "To believe in God is to desire His existence, and what is more to act as if He existed."

Once more, we see in works such as *The Road to Reason*, *Human Destiny*, and *Between Knowing and Believing* the double invitation which makes the permanent rhythm of the philosophical existence: this call of science and of spiritual experience to philosophy.

Mr. Costa de Beauregard has reminded us how we see the commingling or more exactly the union of the biological, scientific problem of evolution, with the spiritual problems of the meaning of life itself; and it is thus that for my part I will repeat once again and in absolute sincerity that history makes us take part in this rebirth of the philosophical exigence which I find so deeply symbolized in the theme chosen by Bazaine to illustrate the last work of Lecomte du Noüy, namely the struggle between Jacob and the Angel Gabriel.

The Future of Man and the Future of Spirit

Dominique Dubarle, O. P.

Allow me to begin this discussion by saying how much I as a Catholic theologian owe to the example and the intellectual effort of Lecomte du Noüy. I remember being received by him during the last years that preceded the war of 1939–1945, his extreme courtesy, and especially his open-mindedness. I remember even better reading his first two books *L'homme devant la Science* and *L'Avenir de l'Esprit*. At that time I had turned to science after having studied philosophy and Catholic theology. I still felt the full force of the challenge offered by the inquiring, scientific spirit to all the first intellectual impediments to which I had been fastened by my religious faith, my kind of life as a religious and a priest. I had already heard Teilhard de Chardin propose the ideal of a symbiosis between the spiritual energy of faith and the equally spiritual energy of science. At times I despaired of ever being able to realize this unity harmoniously at the level of intelligence and reason. In my view, the lyricism of Teilhard de Chardin did not cloak it sufficiently in logical terms. It was then that I met Lecomte du Noüy; and this meeting comforted me by giving me new reason to hope positively for a possible harmony between modern scientific culture and the faith which has been my life.

I found in Lecomte du Noüy, in addition to an unusual modesty and magnanimity, the progression of a spirit who, starting with a consciousness almost devoid of other interests, was in all truth advancing towards what faith represented for me, and even giving a name and Christian character to the spiritual ideal to which his progression was directed. At the outset Lecomte du Noüy did not call himself a believer. His position was more in keeping with the title of a volume of his essays, *Between Knowing and Believing*, published in France with a foreword by Dr. Delauney. But this is precisely what enlightened and helped to convince me. I felt sure that I was not dealing with an apologist who was already sure of the conclusions to which he would arrive and was merely working to set up the mechanical stages by which he would reach them, but with a mind orienting itself in the field of knowledge.

I sensed that Lecomte du Noüy was something more than a man formed strictly by his scientific discipline. For having been so formed and following as closely as he could the rules of this formation he certainly was a scientist. But he was also a man aspiring to and thirsting after the spirit. In this sense he was free; and in exercising this liberty he of his own accord led his thought towards what I believed I possessed but which had created a spiritual difficulty for me when I came face to face with science.

It was in this very liberty, between knowing and believing, that Lecomte du Noüy turned to faith. I admit that it was the experience of this spiritual liberty, stemming from a scientific cast of intelligence, that captivated and sustained me better than any already formed synthesis could have done. For by a kind of equivalence with my own case, I learned from Lecomte du Noüy that faith is a form of liberty and that it can be equally so for science and its spirit. This is perhaps a truth that a Catholic theologian was most in need of and my deep gratitude to Lecomte du Noüy comes from the fact that he gave me the privileged occasion to learn it.

Lecomte du Noüy's conception of evolution suggests consideration of at least two problems that we think are not unconnected. The first of these problems deals with what has already been accomplished in our universe. Life appeared on earth billions of years ago. This is a fact, but what significance must we give to this appearance in comparison to cosmic development considered in its entirety? Can the telefinalist hypothesis of which Lecomte du Noüy speaks be as easily upheld as he seems to think? We may hesitate and feel the necessity to confer again with the astrophysicist and the physicist on this point. The second problem concerns the future and the attempt to achieve a prospective vision of evolution. When he meditates on his origin and his spiritual condition at the present moment of evolution, man considers his own future. But can what is essential in this future really be foreseen? Is the faith in a final fulfillment of the realization of spirit, which Lecomte du Noüy so strongly affirms, sufficiently buttressed by the knowledge of what our humanity has been and by the experience of what we are? Undoubtedly the anthropologist and the philosopher should first be questioned on this point.

I would like to discuss the second of these problems first: the connection that may exist between the real and historical development of man before the present epoch and the ideal hope in a successful progression of the spirit. But after having stated precisely the data bearing on this problem, I shall come back in some sense to the other problem because of the philosophical relationships I see between the two, for certain traits of the adventure of life correspond to this or that characteristic of human development. But while trying to disengage these significant resemblances, I will leave aside a considerable part of the analysis which would be required for the thorough treatment of the problem posed by the transition on our earth of inert matter into living matter.

I would like to point out to those who may undertake to treat this problem in isolation that some clarity could be achieved, at least on the plane of the philosophical interpretation of the facts, from the attention given to other principal articulata of the evolutionary process as a whole.

Lecomte du Noüy, who, as we have said, strongly and repeatedly affirmed his faith in the future of spirit in the end remained very discreet on the way in which he pictured the human march towards this future.[1] In his principal books, *L'Avenir de l'Esprit*, *La Dignité Humaine*, *L'Homme et sa Destinée* [Human Destiny], he seeks to make explicit the moral teachings that can be drawn from this faith. The conjectures relating to the physiognomy, the evolution of the human species beyond its present stage, are merely outlined. The methodical examination of the present condition of the future of humanity has not yet been outlined in these books even though here and there certain important observations are made, such as that on the contraction of the earth which puts the human population elbow to elbow on all the continents.[2]

Lecomte du Noüy therefore hardly tells us in what manner the spirit will tend towards its fulfillment through human development. It will probably occur through the advent of human groups whose spiritual dispositions, acquired in an analogous manner to those resulting in biological mutations,[3] will in the end impose themselves on the collectivities of human kind, while the backward fractions of humanity will eliminate themselves from evolution, left aside by the living growth-forces of humanity.[4] At every moment man leads—in a sense insensibly—towards a beginning of superhumanity, as if evolution in its reality diluted in the human mass the birth of the superman of whom Nietzche has dramatized the image.

Lecomte du Noüy considered this elevation of spirit to be "telefinality" of evolution as a whole. Whether it be dramatic as in the Nietzchean concept, or impalpable, as it proceeds in reality through the daily life of human progress, humanity itself–caught in its concrete existence and progression–is put in the position of being an intermediary result, a minister of evolution in its tendency towards spirit, rather than in the position of being the last goal already assured of essential permanence.[5]

Certain passages in the works of Lecomte du Noüy seem to suggest that the supreme advent of spirit will also be the obliteration of man in spirit. However, the observations which suggest this idea are rather fleeting and it is better not to dwell on them.

The final stage of evolution with its full realization of the spirit is thought to be in the very distant future. Lecomte du Noüy lets himself be guided here by what experiment and science have taught us of evolution, of its great phases and their durations.[6] He tells us that the first expansion of the universe before the appearance of life on earth lasted at least a billion years. From the appearance of life on earth to man, again at least several thousand million years went by. On the other hand, following this scale of duration, the appearance of man, the beginning of the spiritual phase of evolution, is very recent. It goes back a few hundred thousand years. If we admit that the different phases of evolution should be of comparable duration, then we must admit that the universe is still in the very first stages of the dawn of spirit and that the term of the process is still distant. This sentiment of the immensely remote character on the scale of human existence of the last stages of evolution, is also one of the psychological factors which prevents Lecomte du Noüy from giving any outline of the historical path which could lead to it. Faith in the future must be left in a state of faith; the methods of progression which will lead to the accomplishment of the promises still remain quite obscure.

In the meantime, when it is a question of the present and the near future, at the distance of a century or two, Lecomte du Noüy–and this should be underlined–remains rather pessimistic. The last chapter of *Human Destiny* accurately describes the human concentration which has taken place in modern times. It begins with the following sentence: "It is too early to ask all men to think 'universally,' to consider themselves as elements of humanity as a whole."[7]

The spectacle of the heartbreaks, the violence, and the horrors of the Second World War certainly overshadowed the judgment of the author and prevented him from giving himself up to the illusion of an easy natural human progress of uniform growth in every sense desirable. Faith in the future, of spirit, is thus disassociated from the philosophical optimism of the Enlightenment as well as from the scientific optimism of the nineteenth century and of the early years of the present century. Even for the immediate future the previsions of Lecomte du Noüy are filled with misgivings. "It will perhaps be a period in human evolution, a period of anonymous underhanded strife, of distrust against all initiative, a period of regression for true civilization."[8]

The future of spirit can therefore accommodate itself to the human reality of troubled and unhappy historical times in the midst of which spiritual values seem to be degraded and in danger of becoming vitiated. Following Lecomte du Noüy, we will see that the existence of such epochs can themselves be included in the order of things. Be that as it may, faith in the future of spirit is precisely to affirm, even when faced by the possibilities and sad realities of these "somber times," that "Like the ship constantly kept on its course by the pilot who corrects its deviations, humanity may seem to hesitate and waver, however, it will infallibly reach the port which is at the same time its goal and its reason of existence."[9]

From this faith in the final result of evolution, despite all the episodes which apparently go in the opposite direction, Lecomte du Noüy sees in the midst of reality and of history two confirmations of his thought. The first is the success as a whole and "statistically" of evolution up until today.[10] The global fact of the success predominates over the multitude of failures or partial abortions; better still, it envelops in its positive reality this multitude of shadows and distressing episodes. If evolution has succeeded globally until now, why should it not continue to succeed in the same way in the future? The second confirmation is summarized in Christ himself, the highest symbol of all the sublime human emergences. "Christ brings us the proof that the future existence of a superior being, and man's will to be his ancestor are not unrealizable dreams but an accessible ideal."[11] That which was thus realized demonstrates to man the reality of his supreme spiritual capacity.

Our immediate purpose is not to discuss the value of these confirmations that facts give to conviction, but now that the first outline has been given of its contents to come back to a certain number of concepts and to the understanding of Lecomte du Noüy's thesis. For this we will start with what he has written on the subject of the human phase of evolution.

Like all those who have meditated on the place of man in nature or on the way his real being emerges from evolution, Lecomte du Noüy is very conscious of the fact that with man something new supervenes in the rate of the evolutive progress. Man is the living being who has become conscious of his own life, the living being brought forth by evolution and conscious of the development which has begotten him. From now on he is the depositor of the finalities of evolution and conscious of being this depositor. Therefore in him evolution becomes conscious of itself. It takes possession for itself of its incentives and of its goals. Now this consciousness makes man free; free with a liberty which is precisely the very first inauguration of the Spirit. It is the energy which makes of man a being torn asunder from the ascendancy of nature, capable of exerting on it a veritable control. This liberty is not simply liberation from constraints to which other beings, antecedent to man, remain subject; it is, in conscience and by virtue of conscience, the power of choice in the face of alternatives. According to Lecomte du Noüy, the most fundamental alternative is precisely the power of choice in the direction indicated by evolution itself: the choice of the spirit and the opposite choice, the one which maintains his subjugation to the animal condition. Everywhere, under multiple forms, this capacity of choice will be found at the heart of the human drama.

Liberty existed from the very beginning and continues to have its share of responsibility during every moment of human history.[12] It is to it and its unhappy and faulty realizations that we impute the somber aspects of the actual state of humanity and the perils which menace its future. This liberty can bring about the failure of evolution and the abortion of spirit both in the individual and in large human groups.

This requires further explanation. Lecomte du Noüy thus proposes the following idea: at the level of man, liberty prolongs and internalizes for the conscious individual one of the essential characteristics of life in evolution. This is life proceeding gropingly in search of a higher form of existence. Evolution progresses when something is acquired; it makes new trials and prepares the promotion of its successes. The beings it creates are thus subjected to a kind of test of progression. Following the extent of their response in a positive manner, they will become the carriers of future values. Liberty, however, is nothing but the human form of living energy. The position accorded to free initiative in its turn determines the whole of the conditions following which, beginning with man, the progress of the living being can take place. Moreover, freedom is not only a characteristic of the human being; it is his privilege. It is just as much his essential test, a test at the conclusion of which the free being is situated in respect to the supreme finality of evolution.[13]

It is normal that in the case of a test, individuals who are recognized as inapt should be eliminated. In the case of man, it is of himself and freely, by his fault and by choosing animality instead of the spirituality he is called upon to conquer, that the individual eliminates himself from evolution. As in the past, the present of man and the future of our species supposes such eliminations. These are made on the individual basis—at times even on a more collective basis, namely populations—but even more through decidedly failing forms of civilizations, culture and religious spirituality.

The approach to the question we have in mind has now been cleared and we can proceed to formulate it. If liberty in man is the power of conscious choice, if by his conscience man is confronted by the capital alternative of his spiritual vocation and the preference given to his animal nature, if he often chooses to go against his spiritual vocation, cannot this possible choice one day become not only that of certain individuals or of a few particular groups but indeed that of the whole human species? Is it really impossible to contemplate the possibility of a humanity collectively committed to an existence contrary to the spiritual finalities of evolution, or arrested on the margin of its vocation, and having thus reduced to silence the supreme call of the spirit? Should we not, on the contrary, try deliberately to think about this eventuality?

It is certain that Lecomte du Noüy excluded it by his faith in the future of spirit. We do not intend to contradict this faith here but simply to keep it from prematurely excluding the very idea of asking the question which has just been raised. For it would then seem that philosophy would be the loser.

In the *Republic*, Plato evokes this combination of human determinations which are inscribed both in the individual soul and in the reality of collective existence, but in two distinctive modes. He speaks of their being in small characters inside each individual soul and in large characters in the public state of the life of cities. This applies to justice, honor, virtue, and so forth in the citizen and for the state. According to the law of things, there is then a correlation between these spiritual inscriptions in large and in small characters; i.e., that the souls of the just citizens make the justice of the just cities, but also that the injustice of the unjust states can only breed injustice in the souls of the citizens. Now we can seemingly adapt these considerations to our present case.

Indeed we readily recognize, as does Lecomte du Noüy, the individual liberty of man inscribed in "small characters" within such a human soul awakening to conscience. Liberty and its fallibilities at this level are familiar to us. But if this is so, must we not equally consider following the thought given us as a model by Plato and what corresponds to this in "large characters" within the human collectivities and finally the whole species? Would liberty and its essential test concern the individual only? Would it only be proposed for man in small characters? In our opinion the Platonic law is too profound, too essential to be presumed to have been contradicted. Whatever manifests itself in the individual must also possess a reality at the level of the collective being. If in respect to the ends of evolution the individual man is in a position of liberty, then in respect to its ends, humanity as a whole finds itself in a position of liberty.

Indeed the contrary would be almost shocking, for after having recognized this liberty that Lecomte du Noüy admitted as being the principle of all human dignity in the individual, we would end by erasing it in a kind of inevitable fatality at the level of the collectivity. At the same time evolution, which at the level of man creates liberty within the universe, would find itself deprived of liberty. Evolution would continue to be the necessary march towards the finalities of spirit. A blessed march, perhaps, but still automatic and animal in its impetus. Indeed, can we separate conscience and liberty in this way? If evolution is to attain spirit consciously, can it attain it otherwise than freely, and freely up to its own level by dominating and integrating the fate of the human collectivity?

We are thus led to think that the human test inherent in the fact of liberty does not simply act at the level of the isolated individuality or of particular human groups. The totality of the species is itself involved. Consequently if the finality of evolution is really the emergence and the accomplishment of spirit, it is yet not assured of prevailing over the forces eventually brought into play which are contrary to human liberties. At least at the height of philosophical thought one can see no necessary reason to affirm such an infallibility of human destiny. That destiny is spirit, but for the human species, as for the individual, this destiny remains in suspense in the liberty of man. The major crisis of the future of humanity is in all probability still ahead of us.

Before proceeding further we must underline the importance for philosophy of this "time in suspension." With liberty it is not only man who is put to the test but in reality what Lecomte du Noüy calls evolution. The final phase of the latter would not really be an act of freedom without there being the possibility of the most vital disaster. Now this alternative of success or failure which affects evolution attests a change of register when we pass from the universe to the divine reality itself. There where the evolutive transition to the fullness of spirit is made liberty, it is impossible to identify the universe and its nature with God any more than we can make of spirit itself a supreme form of the universe necessarily revealed at the term of its development.

At times Lecomte du Noüy foresees this essential difference of register. It seems to me that he does not succeed in affirming it absolutely clearly nor does he put it systematically to work. His faith in the future of spirit as he professes it destroys, in the answer it brings, the possibility of distinctly wording the question concerning what is in reality the power of evolution. The different tiers in the depths of the human being that such a question presupposes remain hidden from view. To state the question, namely to uncover oneself, is to go to the depths of the being. This does not oblige us to renounce all faith in the future of spirit, but it modifies the initial position.

As far as we can judge, the modification thus perceived is entirely compatible with what the Catholic faith and its theology teach on the subject of human destiny and of its ontological condition. Without meaning to develop this point elaborately, a few brief indications may help us to form an opinion.

The Catholic faith is not content to affirm with vigor the distance between God and the created universe. It insists, moreover, on the incapacity of man to attain by his own natural means the superior finality which is his destiny. The incapacity comes not only from human failure, from the liberty to sin and its consequence. It is a question here of an essential condition of all created reality. Whoever he is, the spontaneous development of the creature does not suffice to make him attain his true and supreme end. To obtain this, some energy coming from God himself (which we call grace) must be united to the personal effort of the being. The failure of creation and that of free man only confirm the fact of this incapacity. They are not the "primary cause."

Let us retranslate these things into the language of Lecomte du Noüy. We must then say that in view of the accomplishment of spirit indicated by the spontaneous outcrop of evolution, the grace emanating from God must necessarily encounter the mounting energies of the universe. For it is from the conjugation of the two that there is born the realization of the ultimate end of the universe. The decisive encounter takes place in man at the level of liberty. If then evolution can succeed absolutely, if its final phase can result in the advent of spirit perceived by Lecomte du Noüy, it is because the liberty of man aided by God has made the desired choice. More profoundly, it is because evolution having become conscious of itself at the level of man has at this very level become a free affirmation of its spontaneous élan. But that man should make the desired choice, that evolution should arrive at this free reconfirmation of itself, is neither fatal nor obligatory. Even in the presence of grace, the liberty in which creation culminates can bring about the defeat of evolution which normally leads this creation to a meeting with God.

Furthermore, we find in the Scriptures indications which lead us to think that if it takes place, the last success of evolution will not be as complete nor as automatic as the faith in the future of spirit postulates in its first generous inspiration. St. Paul tells us that a mystery of iniquity is at work in the bosom of history. In our humanity in fermentation the renegade man is born at the same time as the man responsive to divine grace. The Apocalypse, through its awe-imposing symbolism, evokes a future when the human mass will stand on the side of the liberty which refuses spirit and returns to animality—far more than on the side of the liberty which consents to the true spirit and reunites itself with God. It is possible that these texts seek only to speak of the essentially religious aspects of spiritual decay perfectly compatible, so it seems, with many human indications of a growth of spirit. But these texts are nevertheless there to strike us with a certain radical concern in respect to the destiny of our own species and to keep us from a certitude too easily acquired on the subject of the favorable outcome which would in any case be reserved for it. At stake in the eventuality of somber periods which Lecomte du Noüy does not exclude, there is—more than he wishes to admit—the risk of a total perdition of the human species, by counting only on itself and on the spontaneous powers of evolution.

From the present point of view these theological notations have only been brought up as an external confirmation of the question evoked by reading Lecomte du Noüy. But there is another type of confirmation which derives from the first of the two themes mentioned in the beginning; namely, that of the place apparently given to life in the whole of cosmic reality.

Current science reveals a universe so immense that it confounds the imagination. With the help of modern instruments, the visual radius of astronomy reaches to the confines of a sphere of some eight billion light years. Numerous tens of billions of galaxies fairly similar to ours are scattered inside this sphere. On the average they are distant one from another by about two million light years. Our galaxy contains around one hundred billion stars, many of which have characteristics closely resembling those of our sun. When compared to this vast whole, our earth has not even the dimensions of an infinitesimal speck of dust. And yet, until now at least, the only known realization of life is on this earth.

Is it possible that one day we shall convince ourselves that life exists elsewhere? Planets can have been formed around suns similar to ours. We cannot eliminate the possibility that in one or another of them, or even perhaps in billions of them, the adventure of life can have been repeated. However, even in that case two things would remain true. Only an infinitely small part of the enormous mass of the universe would in fact attain a living condition. This accession, moreover, would have been produced in localities so distant from one another that there would be little hope of ever seeing life joined from one star to the other and forming through space a true community of cosmic vitality. The universe makes the realization of life quasi-infinitesimal and isolated at its own level.

Such is the modern aspect of our understanding of things, so different from that of the ancient cosmos which, at least at the start, pictured the entire sky as the abode of the highest life and all the stars as an army of immortal living beings established in the fullness of a spiritual existence. This modern view is enigmatic. In truth, it should astonish and arouse meditation more than it does.

In any case, intellectual astonishment seems inevitable when one wishes to follow Lecomte du Noüy's concept of evolution. If life is really the finality of evolution and of the formation it accomplishes in universal matter, how is it that things are disposed in such a way at the start that the cosmic realization only attains the actuality of life so exceptionally? So many physicochemical conditions must be united for us to find ourselves living here in an acceptable biological environment (and it is so rare for them all to be present) that we hardly dare to say that the universe was formed for this end. It seems more probable that the universe of expanding matter realizes a little life into the bargain, in a highly exceptional and almost indifferent manner compared to the massive vastness of the physical substratum.

I know that Lecomte du Noüy will then speak of a distant finality, of a "telefinality" rather than a finality. For his part, Teilhard de Chardin will try to give a certain comprehensiveness to the emergence of life from inert matter by defining, aside from the spatial and temporal magnitudes, an axis of complexification of reality as one of the essential dimensions of nature. All this is important and should be welcomed. Gradations brought to the concept of finality and arguments taken from the structural powers immanent in nature show us that if the position of intelligence is difficult it is all the same not completely desperate. But the fact remains that very little of the universe passes into life to reestablish itself at the level of the superior and distant finalities assigned to its evolution. Very little, and at the price of an immense number of beings which must then be considered as fixed on the deadlock of the cosmic process—expanses, stars, floods of errant particles.

Now this repeats itself, with perhaps a little less disproportion, at the ulterior level of the rise of beings. There is a great deal of life on earth which does not attain to intelligent conscience. One unique species achieves it: ours. At the moment when spirit itself comes into question, Lecomte du Noüy is the first to recognize the truth contained in the ancient saying that true humanity lives in very few individuals–*Paucis humanum genus vivit.* A rather somber law seems to govern the plan of telefinality; namely, that in the great competition between beings in pursuit of it, almost the entire mass is brushed aside and kept from obtaining it. Only a very small remainder finally gains access to it.

Here we rediscover at the level of the human being (of the collectivity and of the future which can be envisaged for it) the fact of liberty and the question specifically attached to it. For Lecomte du Noüy, liberty organizes itself entirely in man as a function of the lived alternative of his consent to the spiritual aim inscribed in evolution or of his rejection or betrayal of this aim. But precisely, it is liberty, namely (also in its execution) the highest realization at all levels of the superior conquest of existence. Can it then avoid reproducing in the bosom of the human species that which shows itself with a kind of constancy outside it? Can then the highest finality, the full life of the spirit, be anything else as long as this universe lasts, but an exceptional realization far from becoming the common form of all existence? On the basis of simple philosophical reflection, one is tempted to answer these questions by stating that the spirit runs the risk of being hardly more for the future of human populations than man himself has been for the future of the living populations of the animal realm. The vocation of spirit haunts the human gender, as no doubt the vocation of man haunted the animal realm towards the end of the tertiary period. But who of our race can vaunt himself as being completely born to spirit?

The birth of liberty within the human being necessitates putting the infallibility of evolution in question metaphysically. However, the considerations we have just made concerning the conditions of this appearance of life in the universe lead to another more cosmological form of this same question. This is the relationship between our two problems. Evolution in each of its great stages shows a certain success which is never global. The part which succeeds is rendered strangely small and local by the very severe conditions of partition. These conditions oblige one to meditate every time on the small number of the elect. This is what gives to the cosmic process its very particular physiognomy which is at the same time enigmatic and significant. In respect to the nobler states of beings, our universe is less well disposed than we would like to imagine it.

The pessimism of these considerations must, however, be corrected. Just now I suggested that the "law of division" between success and failure seems to become less severe in proportion to the elevation attained in the scale of beings. If very little matter reaches a living condition, there is perhaps (proportionately speaking) a little more of the human being which raises itself to the authentically spiritual level. But above all we must add that the higher one rises in the scale of beings, the greater is the positive repercussion possessed by the conquered energy over the mass this conquest at first leaves behind.

Confined to the terrestrial locality where it was born, life can to all intents and purposes do almost nothing for the immense universe which elsewhere has remained powerless to live. Man acts more upon the mass of living matter which he humanizes to a certain degree; the spirit which appears in man does not cease to leave a fertile wake in the midst of all humanity. From this point of view the latter indeed accedes to a certain condition of spirituality—no doubt derived, but real—of which we can today measure the collective amplitude and continue to inquire into the promises of the future. By its communicative power the spirit when conquered in an individual achieves in the bosom of the human group an ulterior victory over the exiguity of its initial success. However, no matter how substantial this ulterior victory can be, one can ask oneself if for any of its beneficiaries it can be equivalent to the act of original reestablishment which alone can be carried forward by the ardor of its desire; the initiator of the wake of spirit has lived for his sake. We must graduate in proportion the faith in the future of spirit that was professed by Lecomte du Noüy. For my part I would distinguish between a future of spirit, of which no philosophy can tell me with certainty who amongst us today is the bearer, and the future of man, which appears to me born every instant—not in coincidence with the future of spirit which still seeks itself in the development of man, but as the origin of spirit which has already appeared amongst us at this moment. There is here a slight difference. But I only formulate it whilst expressing to Lecomte du Noüy the gratitude of an intelligence to whom he gave the very possibility of formulating it.

You will allow me to end by another very short theological note whose first premise I think Lecomte du Noüy would have accepted. Following Lecomte du Noüy we wanted to consider the human development as being the phase of evolution which can advance towards spirit. Taking into account the present remarks, should not a Christian consider Jesus Christ as the unique being in the midst of humanity in whom was accomplished from the outset the supreme completeness of this advance? In this sense, we would have to say that evolution has already achieved in our midst the singular success of which it was capable. Our future as men depends on the free play of communication with the mass of energy which rested on the one whom the Christian faith affirms is the first personage of our species and of all times to be fully man and Son of God.

I have no more to say, and I leave this Christian interpretation to your judgment.

NOTES

1. Pierre Lecomte du Noüy, *L'Avenir de l'Esprit* (Paris: Gallimard, 1949), p. 306; *La Dignité Humaine* (Paris: Brentano, Editions du Champ de Mars, 1944) p. 242; *Human Destiny* (New York: Longmans, Green and Co., 1947), p. ii.
2. *La Dignité Humaine*, p. 239; *Human Destiny*, pp. 259–262.
3. *La Dignité Humaine*, p. 208; *Human Destiny*, pp. 225–227.
4. *Human Destiny:* "They whose souls have been perfected in the course of their passage through their bodies, who have fully understood the conflict between the flesh and the spirit, of which they have been the stage, and who have triumphed over matter; they alone represent the evolutive group and are the forerunners of the superior race which is to come" (p. 187).
5. *La Dignité Humaine:* "From the beginning, the final goal to be reached was not the human form but conscience and spirit" (p. 21).
6. *L'Avenir de l'Esprit*, pp. 271–276; *La Dignité Humaine*, p. 205.
7. *Human Destiny*, p. 258.
8. *Ibid., pp.* 263–264.
9. *Ibid.*, p. 266.
10. *La Dignité Humaine:* "From a statistical point of view evolution has succeeded" (p. 164).
11. *Ibid.*, p. 161.
12. Lecomte du Noüy tries to discover it at the level of the origin of humanity by making in his own way the exegesis of the second version of the Creation in the recital of the fall in Genesis.
13. *Human Destiny:* "Liberty is not only a privilege, it is a test" (p. 118).

Science and Religion

Paul Weiss

Lecomte du Noüy and Teilhard de Chardin were similar in at least three basic ways. Both were distinguished scientists with strong convictions that scientific inquiry yields reliable truth on which to build large-scale systems. Both were practicing Roman Catholics, though one was a layman who came to semi-orthodox religious convictions rather late in life, while the other was a dedicated religious man who often seemed to verge on heresy. And both offered arresting, not altogether incompatible ways for reconciling science and religion, and incidentally, secular with religious history, and ethics with revelation.

Both men have made the intellectual world aware of the need to accommodate religion as well as science; they have alerted us to the possibility of treating both enterprises as part of a single cosmic totality in which the facts certified by the one are allied to the faith characteristic of the other. Their views have had great influence; but they have defects too. An examination of the main alternative approaches that can be made to the question of the relation of science and religion will, I think, make both the strength and weakness of their positions a little more evident.

I

It is possible to begin with a view which stops at once. We do this when we claim that science and religion are incompatible, that they are intrinsically and consequently forever irreconcilable. In a religious era, one who takes this approach will be inclined to dismiss science as foolish, and perhaps even as evil. In a scientific epoch one will tend to attack religion as unintelligible, and perhaps also as perverse. In a more ecumenical period they might be said to be equally desirable but nevertheless still incompatible in attitude, method, result, and value.

It has been a long time since religiously oriented thinkers were able to persuade themselves and others that science could be dismissed because its claims conflicted with those urged by religion. It is more common to find religious assertions shrugged aside because they are taken to be not in consonance with the scientific temper and discoveries that it is common to acknowledge today.

Both rejections suffer from the same fault. They both are question-begging, since they suppose that only one of the two positions provides a test of what is legitimate or acceptable, despite the fact that its terms and values are rejected by the other.

The ecumenical version of the view that science and religion are irreconcilable avoids the defects of the other versions. It also has had the strong backing of Wittgenstein. According to it, both enterprises are equally legitimate. Each is a "language game" worth studying by philosophers. Neither game is superior to the other, each is to be taken as it appears; each has its own vocabulary and grammar, and the implications of both are to be pursued.

Neither Wittgenstein nor his followers has accepted this position with full seriousness, for they do not give full weight to all languages or 'games'. Theirs is a program not yet completely illustrated in performance. Mysticism, metaphysics, and art have received hardly any consideration by them, in good part because common sense or science has been tacitly assumed to have a preferential status.

Even at its most generous, the position is in difficulty. If all one does is to study a language, one will not know if there is anything answering to it in fact; nor will one have any way of determining whether or not a given language is doing justice to its own subject matter. All one can do is describe, and perhaps clarify and make a little more consistent what men in fact are saying.

A little advance is made if one gives up the attempt to stay only with languages and attends instead to the different entities with which different disciplines deal. If science and religion are irreconcilable our world must be radically pluralistic, not comprehended in a single overall view. But this position, like the others, necessarily makes an assumption which in effect denies its own finality. To know that both science and religion exist one must be able to take a position encompassing both; to know that there are items for each, one must be able to look at both. Just so far, the two outlooks are not incompatible. They may diverge and even have functions and assertions which conflict with one another, but they will nevertheless be somehow together. Not until one examines the nature of the perspective which embraces them both, will one know whether or not there are objects subtending both of them and in which they are concretely reconciled.

II

Common sense observes carelessly, and reasons without subtlety. Too often it speaks obscurely and inconsistently. What it knows is rapidly overlaid with false beliefs, practical demands, and arbitrary conventions. Nevertheless, it is not altogether without merit. The most advanced science can be viewed as a refinement of much of what common sense has painfully but surely mastered. The same commonsense root can be claimed for religion. But it can be well argued that religion fails to advance as far as or as surely as science does. From this perspective science is seen to do somewhat better what religion is also taken to do—to make sense of what we experience or know as commonsense men.

The Bible, it has sometimes been said, offers reliable clues which trained historians and archaeologists can pursue with their controlled methods and sharpened instruments. Some believe that biblical dietary rules anticipate, in a rough form, the clear and well established dictates of modern medicine. Religion, it is thought, blindly sees what science will clearly know.

A variant of this position is offered by Comte, who supposes that there was once an age of religion which gave way to a superior age of science in which we now fortunately live. Another variant is to be found in Peirce who takes religion to be a poor way and science to be the best way of making our ideas clear.

This position has a counterfoil. There are some for whom science is religion in a minor key, or at worst religion secularized and therefore distorted. Something like this view was adopted by some of the clergy who opposed Darwin and Huxley. The striking predictions of science, the multiple confirmations that some of its claims have received on the part of many independent investigators, and its great success in promoting technology and industry have made this position seem not very plausible, though it is in fact no less reasonable and defensible than its opposite.

The defenders of neither of these positions seem to understand the power or justification of the other. Neither sees that their different tasks and objectives permit them to be on a footing, not as irreconcilable but as supplementary agents to be employed in a single search for truth. They both overlook the fact that they both may be mistaken. Neither science nor religion may be reliable. And, it is not to be forgotten, there are other enterprises which claim to be even more basic than either. History, art, and metaphysics—to name only some—insist that neither science nor religion, nor both together give us all that we need to know. The war between science and religion is to them only a skirmish for limited territory. The observation suffices to make one hesitate to accept the view that either science or religion is a primary enterprise which the other expresses in a truncated or distorted form.

A stronger position than any of the foregoing, but one not altogether alien to their spirit, takes both science and religion to be somewhat equal in respectability while supposing that one of them has a reach which allows it to take the other as a special case. Dewey, for example, acknowledges many legitimate enterprises, but treats them all as the proper topics of a scientific inquiry. In somewhat his spirit, there are those who speak of religion as teaching the basic virtues of honesty, courage, selflessness, and objectivity which the pursuit of science requires. Some scientists, as Du Noüy has observed, say that the more they contemplate nature the surer they are that there is a God. Their science, it would seem, is a subdivision within a religious outlook.

Neither the view that science is enhanced by being given a religious context, nor the view that religion is to be understood from a scientific perspective, need deny weight or dignity to the contained enterprise. But neither grants that this contained enterprise is as full or as basic or as meaningful a subject as that which encompasses it. Once again, though in a milder form, the evaluations and outlook of one view are allowed to obscure the merits of the other.

III

An extension of the preceding view takes science to affect every part of religion. A proper understanding of religious (or any other) phenomena can then be thought to depend on assuming a scientific perspective. Here religion is given its own place but only so far as this is seen to be within the nature that science knows and masters. I know of no one who has made this position at once plausible and clear. The opposite stress, however, which takes every occurrence to be in root a religious phenomenon, to be understood by seeing it as a more or less faint or accurate image of God, has been brilliantly adumbrated by Bonaventura. Tillich can be said to have given this approach a modern form. For thinkers such as these, true history is religious history, true courage is religious courage, true conduct is religious conduct, and whatever occurs is God in a diluted and limited form.

Once again we come to the repeated objection: the view that one discipline affects all the others sustains two distinct positions, each of which minimizes one of the disciplines. We need a view which does not have these defects. This holds that the worlds of science and religion are both phases or aspects of a third world. This position has two main forms, one exploited by emergentists, the other by substantialists.

Emergentists conceive of a hierarchy of realities all having a common base and issuing one from the other to yield a series of parallel planes, all equally real. For them there is a world of electrons and another of molecules, on top of which there are those of cells, the lower living beings, man, and eventually God. When the levels are treated as not only higher above the base, but later in time and superior in value as well, we approximate the position of Du Noüy.

Lecomte du Noüy is an evolutionary emergentist. The universe, as he sees it, is progressing toward the achievement of more and more freedom. At the present stage we have man, the freest of all living beings; eventually that man will be perfected; he will come closer and closer to having a conscience and virtues which are like those of that perfect being, Jesus Christ.

The view is not as well integrated as one would wish. Lecomte du Noüy allows a rather naive physiologism to intrude: "It is not the eye which sees, but the brain" (*Human Destiny*, p. 15). "Our reasoning faculties, i.e., our brain cells . . ." *ibid.*, p. 17). From here he constantly slips into an uncritical subjectivism. Again and again he says such things as "This picture is construed *by the mind*, and that as a consequence it is dependent on the structure of the brain, on the sense system which puts us in contact with the outside world, and on the logical mechanisms which are at the base of the interpretation of direct sensorial observation" (*ibid.*, p. 15). Or, "Phenomena only exist as such in our brains" (*ibid.*, p. 17); or, "The electron . . . which is essentially a creation of our brain" (*Between Knowing and Believing*, p. 215).

He readily grants that physics and chemistry are the masters of the inanimate world where the laws of chance hold sway. But he thinks that life itself cannot be explained by physics or chemistry or by any of the laws knowable by them. "It is totally *impossible*," he claims, "to account scientifically for all phenomena pertaining to Life, its development and progressive evolution . . . unless the foundations of modern science are overthrown" (*Human Destiny*, p. 37). Again, "The laws of chance only express an admirable subjective interpretation of certain inorganic phenomena and of their evolution" (*ibid.*). And again, "Using the same methods which have been proved so useful for the interpretation of the inanimate world, it was impossible to explain, or to account for, not only the birth of life but even the appearance of the substances which seem to be required to build life . . ." (*ibid.*, p. 38). Again, "With the advent of death natural evolution has striven to evade the statistical hold which dominates the inorganic universe and has prepared the way for the advent of human liberty" (*ibid.*, p. 53). And, "Man escapes from the grasp of the physicochemical and biological laws" (*ibid.*, p. 93). And, "Natural evolution, which unfolds in a direction *forbidden* by science, i.e. toward more and more improbable states . . ." (*ibid.*, p. 137). And finally, most emphatically: "*The sciences of inorganic matter are indispensable but insufficient for the study of living matter, for neither chemistry nor physics can shed light on the harmonious coordination and subordination of vital phenomena*" (*ibid.*, p. 39).

But he is no simple vitalist. Instead he envisions a single, cosmic, spiritual, final cause which operates on living phenomena to bring about greater and greater freedom. The terminus of the cosmic venture is for him a world apparently only of Christians—free, devout, and moral. The position prompts him to make some rather individualistic interpretations of the Bible. He thinks, for example, that it supports the view that man's body evolved and that he, as a being with a conscience who can know and do good and evil, must be a divine product. I find little of this to be plausible or well defended. If it were put aside, the main tenor of his view would not yet, I think, be seriously affected. It is possible to hold to a continuity of kinds which starts with submicroscopic entities and points toward perfected man, and perhaps beyond, without having to make a physical and chemical interpretation of whatever is inanimate, or to embrace a physiologism, or a subjectivism. One will still, though, be left with some serious perplexities.

What is the nature of that evolutionary process which surges through all that lives? What is that finality which guides and perhaps pulls some beings, such as man, forward? How and why does God add something to man's body to make him a being with a distinctive destiny? What warrant is there for the acceptance of the views of but one of the world's major religions? In the absence of clear clues to the answers to these difficult questions, it is wise to restrict our field further and attend only to a minimal temporal emergentism, for this apparently would remain no matter how those questions were answered. But I think this, too, cannot withstand critical scrutiny.

There appears to be no evidence that the outlooks of science and religion, or the realities and experience which these acknowledge, are in an order of early to later, or lesser to better. Putting aside the fact that for many a responsible thinker subatomic particles have no genuine reality, there seems to be no evidence that there was once a time when there were no complex inanimate beings or there was no subhuman life. And if it be plausible to treat biology as a discipline concerned with laws which work in opposition to those expressed by physics and chemistry, what should deter us from following this line and claiming that psychology, sociology, and religion also defy the trends of biology? We would then still have a "telefinalism," but it would be one which extended far beyond the biological realm.

IV

Substantialism, I think, offers a better way to reconcile science and religion than is possible to an emergentism, static or temporal. We cannot deal with it properly, however, until we are alert to the fact that there are other sciences besides physics, chemistry, and biology, and other religions besides Christianity. But we will still have left whole areas of knowledge and being untouched. We must go on to acknowledge other enterprises, such as history, art, and politics. These seem to be as distinctive and as fundamental as our chosen two.

Substantialism has two main forms. In one a reference is made to spatiotemporal substances, abstractions, and refinements of which are the special topic of different inquiries; in the other, one attends instead to ultimate realities on which the different inquiries depend and which they explore under limiting conditions. They are not mutually exclusive.

Spatiotemporal substances are commonsense objects, purged of their irrelevancies, conventional additions, arbitrary categorizations, and pragmatic interpretations. Science can be said to deal with such substances as subject to a cosmologically relevant mathematics, and to be analyzed into smaller particles all of which, when treated physically, have a place in a repetitive time. Religion can be said to deal with those very same substances but only so far as they have been given a sacramental role by virtue of God's presence in them. These substances also ground a history so far as they are taken to be inseparable from their futures and to be related to their pasts within a humanized context. Psychology, sociology, and politics all occupy themselves with some of these substances—men and what is related to them. These and other disciplines can all be viewed as originating out of an attempt to deal clearly with distinct facets of commonsense objects. All of them together offer an articulation of the nature of spatiotemporal substances, i.e., of purged commonsense objects.

In these substances the import of the different facets is preserved. This is consistent with saying either that the different facets are indifferent to one another or that each encompasses all the others without compromising their independence. The former is the more cautious position; the latter is bolder and has been debated for centuries in a very special guise. The latter has bearing on the problem of incarnation.

Theologians have long debated the question of how an incarnation is possible. In the West it has a sharpened form in the question of how Jesus Christ could be both human and divine. The orthodox position can be summarized in the view that he was wholly man and wholly God. So far, the position sounds as though the two dimensions existed side by side. But the body was lived in no simple carnal way; and the divine nature was given a bodily location. It was thought Christ's bodily existence gave his divinity a new import; Jesus's divinity made his body glow with virtues all others can only approximate. Such an account provides a paradigm for the doctrine that in spatiotemporal substances distinguishable facets interpenetrate and condition one another. I think that view is sound.

But whether we say that the different facets of substances interpenetrate or are merely coordinate, we are faced with the question of how the substances are known. A good case can be made for the theory that art is a means by which one grasps the nature of substances as undivided into the facets which are studied in the different intellectual and practical disciplines. That possibility suffices to point up the fact that substances are to be known in ways and through an activity distinct from those which deal with their facets.

This approach has limitations. It leads to the consideration of interpenetrating localized factors without making provision for the knowledge of how they can be together. It does not tell us of the principles and laws which the facets illustrate, nor of any beings on which they may depend or toward which they might point.

The sciences do not restrict themselves to a study of formalized and abstract facets of substances; they also formulate theories purporting to tell us of the nature of the entire space-time-dynamic cosmos and the kind of entities to be found in every part of it. Religions do not merely present us with sacramentalized objects; they also try to bring men into closer contact with God who exists outside the universe of space-time things.

We cannot stop with a consideration of finite substances with interpenetrating facets. We must acknowledge realities beyond them as well. We will then come to the position that there are a number of ultimate realities to which the facets of substances and the substances themselves refer to as ground or correlatives. Science, religion, and ethics can be understood not merely to deal with abstract dimensions of substances but with such substances as qualified by transcendental realities—the whole of extended existence, God, and prescriptive possibility. Substances have a mutually infective set of distinct facets, those facets are qualifications that are partly determined and sustained by transcendental realities from which they are inseparable. Such an account completes the view that there are spatiotemporal-dynamic substances, aspects of which are the topics of special disciplines. The doctrine of incarnation and the more general hypothesis about facets of substances which it prompts are evidently consistent with this acknowledgement of transcendental realities.

A view such as this is left with the problem of the way in which the transcendental realities are together. This is a metaphysical question to be dealt with through the use of the characteristic metaphysical methods of dialectic, systemization, and self-criticism. No matter how one succeeds in this adventure, one will have passed beyond the point where the reconciliation of science and religion is a problem, at the same time that one will have avoided either exaggerating the significance of either, restricting their scope, overlooking other disciplines, or making the error of taking a genuinely inquiring mind to live in anything smaller than the entire cosmos. In summary, I think we must say that the world contains a plurality of substantial beings which are affected by transcendental realities to turn them, among other things, into entities in a cosmos, or into entities which have a sacral status, the one being studied in science and the other accepted in religion.

Lecomte du Noüy had a cosmological temper and a strong conviction that science and religion could be reconciled without injury to their basic intent. It is somewhat in that spirit that a solution alternative to his is here being presented.

Deity as the Inclusive Transcendence

Charles Hartshorne

Professor Weiss, whose work has been a constant reminder of the full scope of philosophical concerns, suggests that to do justice to both science and religion, we need to accept the view that there are spatiotemporal substances with interpenetrating facets, these facets being partly determined and sustained by transcendental realities, one of which is God. Science studies the substances as parts of a dynamic cosmos, and the religions present us with sacramentalized objects and try to bring us closer to God.

There is grandeur in this view. It is big enough—and perhaps I may add vague enough—to accommodate Lecomte du Noüy, Teilhard de Chardin, and many others. The vagueness is not, I think, entirely overcome by taking into account Weiss's relevant books, especially *Modes of Being*. My own personal conviction is that the transcendent reality of God, properly conceived, is the entirety of the transcendental, and that the duality of God and his creatures is the whole story. This of course is a classical position. However, there are various ways of conceiving the duality; and I am no more persuaded than Weiss that the usual formulations are adequate. It is another question how the defects of these formulations are to be remedied. Is it by adding to deity two other transcendent realities: *existence*, or the whole formed by spatiotemporal substances; and *prescriptive possibility*—called also Ideality in Modes? I cannot think so. I shall try to sketch my own view of what is needed.

What Weiss calls substances, individuals enduring through time and extended in space, are somehow provided for in most metaphysical systems. The use or avoidance of the word *substance* is secondary. Some of us find the historical associations of the word as a technical term misleading. And there is a division of opinion as to how the enduring individualities are related to concrete momentary "states," events, or instances of becoming. "Process philosophies" take the most concrete or determinate units of reality to be not things or persons passing through successive changes, but momentary actualities which become or are created, though they do not change. Change is the successive becoming of distinguishable—though often intimately related and closely similar—actualities.

This has for two millennia been the Buddhist view. It was also Whitehead's, and, with some hesitation, Peirce's. I believe it is correct, though extremely easy to spoil by one-sided or crude formulations. Formulated with subtlety and balance, doing full justice to the intimate relationships mentioned, it does not prevent us from admitting that there are enduring individual realities, call them substances if you will. Of course there are persons, other animals, and many other kinds of persisting individual entities. Not many of us deny this. It is not the reality of "genetic identity" (identity through change) that is moot, but it is the logical structure or status of this identity. The process view is that the identity is somewhat abstract, and that this abstract reality, like all abstractions, must be defined through the more concrete, momentary actualities and their qualities and relations. Genetic identity is, after all, a relation, and it is not the relation of logical or absolute identity, for it is tolerant of a large measure of difference. It is not $A = A$. Rather, it is a relation of successive and partly contrasting realities to one another as forming a certain kind of sequence. The sequence is not a class identified by its members, for the same individual could have had a partly different past, and his present identity as that individual does not, in spite of Leibniz and others, entail the details of his future career.

Beyond ordinary spatiotemporal individuals, and the events or states constituting their full actuality, what else is there? There is, I agree with Weiss, the transcendent or eminent individual, God. I am not sure that Weiss applies the term "individual" to God, and of course Paul Tillich, by implication at least, denies its applicability. The issue is at least partly verbal. True enough, God is not "one more individual being," since it is other individuals that are added to the primordial being, rather than vice versa. Nor is it very appropriate to speak of God as "*a* being," for he is *the* one and only possible universally presupposed being, individual in a radically unique fashion. Still, God is individual, not a mere universal or a mere kind of being. He is the *universal individual*, unique in both aspects. Here alone these normally contrasting ideas can coincide. This is saved from contradiction by the fact that in this philosophy, individuality—even in ordinary cases—is somewhat abstract or general compared to the momentary states in which alone it is fully concretized or particularized. Only in deity is the relevance of an individuality to diverse times and places (as my self-identity has functioned in many parts of the world, and through seventy years) an absolutely universal relevance or efficacy. Weiss's view of God seems to have at least some affinity to this.

"Possibility," in my view, is but an aspect of existing individuals, or finally of their momentary actualities. These actualities, I have said, become or are created. They are not caused in the classical or necessitarian sense according to which what happens is the only thing which in that situation causally could have happened. A paradigm of a created actuality is a momentary human experience. An experience is an emergent or partly free synthesis of antecedent actualities which are its data. An experience must have data, and these cannot be provided by contemporary or future events. The synthesis of the data into a new actuality must be at least trivially free, for the data could not dictate their own unification into a new actuality. This would be contradictory or meaningless. The new unity must be new; it must be a creation. But it is not a creation out of nothing. The data have to be expressed in the unitary character of the new actuality. This relation is causality. It is not classical, deterministic causality, but allows and even demands an aspect of indeterminacy. The admission in current science, that—at least for all we could know—causal regularities or laws are only approximate or statistical, makes it easier today than it was in the past to accept the creationist doctrine. However, Peirce and some others held such a view well before quantum mechanics.

What, then, is possibility? It is the creative leap between data and the new synthesis. Possibility is the partial ambiguity or indefiniteness of the future as in the present, or of that which not only can but must be further determined subsequently. This is the essential meaning of "possible." All more special meanings can, I hold, be derived from this one. Possibility is real so far as futurity is real. But futurity is only an aspect of actualities as such, as events. They have data whose futures they further determine, and they will themselves be data in actualities which are foreshadowed, but not fully defined, by their own futures. Apart from pastness and futurity, actuality is nothing. We need not add possibility to actuality; we already have it in the irreducible idea of process or actuality.

According to Weiss, God and possibility require each other. In a sense I agree. But this is nothing additional to the idea of God as eminent individual, concretely actualized in an eminent form of actuality, one aspect of which is eminent possibility or futurity. "Eminence" can be defined as "unsurpassability by another," the last two words indicating that self-surpassing is compatible with eminence. It must indeed be the eminent form of self-surpassing. And there must also be an eminent form of "unsurpassability even by self." For so far as an individual must always be at least itself, whatever defines this identity can never change or increase. But since genetic or individual identity is abstract, it is compatible with increase in more concrete aspects. There must be divine potentiality; for individuality, actuality, and possibility are inseparable. We have to consider each universal or categorial concept as dividing a priori into the eminent or divine form and the ordinary or nondivine form. There is eminent and noneminent actuality, and eminent and noneminent potentiality, and this twofold duality is a priori not empirical. The correlation God-World, taken in its most abstract sense, is subject to an ontological proof. It cannot be unactualized. But only the eminent side of the correlation indicates an individual.

What Weiss calls existence, or the whole of actuality, is included in the divine actuality. Only divine actuality includes all actuality, and likewise only divine potentiality includes all possibility. God actually has the de facto universe in his own awareness, and whatever universe could exist he could and would have it. Thus divine actuality and potentiality are definitive of actuality and possibility as such. This has some analogy with Tillich's "God is being itself."

It is perhaps needful to remark that in a process metaphysics "the whole," or "the universe," or "all existence" cannot refer to a totality fixed once for all, but are demonstrative like "here" or "now," gaining additional or new meanings with each use. The world is progressively created and God acquires each additional actuality as "enrichment" of his own life. At this point Berdyaev, Fechner, Lequier, Whitehead, and still others agree very well. For these men God is indeed a living God.

I have said almost nothing about the special biological and anthropological problems dealt with by Lecomte du Noüy or Teilhard de Chardin. I believe with the latter that there is no such thing as mere dead matter; that mind, in some humble form at least, is everywhere. I also believe, with both men, that the laws accessible to physics, with its present methods, are inadequate to do justice to biology and physics. Wigner and Heisenberg, among physicists, have hinted as much. The adequate science of nature, if we ever have such a thing, will not treat the presence of human thought or even of animal feeling as irrelevant, or a mere epiphenomenon, as present-day physics does. The unity of a vertebrate animal, in particular, as feeling and remembering and so on, must make some difference, however slight, in the movement of the molecules in that animal. The formulas of quantum mechanics make no allowance for such a difference. To that extent they cannot be universally and absolutely valid.

I have one basic conviction, however, which some scientists with religious beliefs may not entirely like: the essential nature and existence of deity are not subjects for empirical argument. By "empirical" I mean, as Popper does, conceivably falsifiable by experience. I hold that no conceivable experience could imply the nonexistence of God. For God must have eminent capacity to coexist with any universe and any experience that is possible. To speak of a possible nonexistence of God is to suppose that divine potentiality is *not* the measure of possibility as such. But to be that measure is inherent in the idea of deity.

Am I denying that "the heavens declare the glory of God"? No, for to the believer they do declare it—but with the understanding that any conceivable heavens or absence of heavens would do so for a sufficiently wise observer. The eminent cause must be in any effect; if it is not seen there, so much the worse for the penetration of the seer. The special character of this actual world, which is what science seeks to reveal to us, shows us not that God exists (for he would do so in any possible case) but only what universe he contingently has. Apart from science we could know that God, the eminent one, exists; but what sort of world he has we could not know. Thus it is contingent truths about God, beyond his necessary bare existence and self identity, that are the province of science. Moreover, apart from his contingent aspects, deity remains an empty abstraction. Knowing God in this abstract aspect alone, we could know that he surveys all reality, but not what specific or individual realities the survey encounters. We could know that he orders reality with eminent wisdom (assigning to the creatures appropriate degrees of scope for their own self-determination), but what the resulting order *is* would be hidden from us. We could know that God's deeds were right, but not what he has rightly done. And so we could not know our role as his servants in his cosmos. To illuminate and enlarge that role is the sufficient religious contribution of science.

From my point of view Weiss oversimplifies the problem of changing yet identical individuals, and overcomplicates the problem of transcendence. From his point of view, I suppose, I overcomplicate—he perhaps thinks I deny—individuality; and I oversimplify transcendence. He finds a lack of unity or integration in my individuals, and I find a lack of unity in his entire system. Reality nowhere seems to add up in it. In my view, all abstractions and possibilities are contained in concrete actuality, all past actuality in present actuality, and all ordinary actuality in divine actuality. Thus quite literally all reality is in God. Yet both God and every other individual have some creative freedom. Nor is God a mere impersonal principle, as in classical pantheism. He is an individual, eminently acting upon and eminently receiving influences from the nondivine individuals. The "glory of God" is neither God apart from the world, nor the world and God, but the world taken into the divine life. And this life is genuinely such. It has a settled past and a future open to endless further enrichment. That it is not already maximally rich is no defect; for the concept of maximal richness is either contradictory or meaningless. God cannot have as actual all possible values; for possibility, futurity, is inexhaustible, and at every moment some otherwise possible values are rejected once for all. God can have me choosing a career as a philosopher, or he could have had me taking another choice; not even *he* can have both. And if it be said that it does not matter to God what we choose, I can only say, "So much the worse for that way of thinking about God."

To much of the foregoing there may be analogues in Weiss's wonderfully complicated view of transcendence. To a good deal of what he says about religion I feel profoundly sympathetic. Religion, he indicates, is more than mere theory or knowledge. It brings us closer to God and makes objects glow with reflected divine light. It creates in us not merely new ideas but new characters. It transforms what the Gospels call the heart of a man. That is why it employs rituals and sacraments. It must move the entire personality, as mere concepts cannot do. It is not mere morality but "the morale in morality," as a friend has said. Weiss's *The God We Seek* (perhaps his best book) is a fine study of the major religions as various manifestations of man's religious quest. Religion is the crowning phase of man's self-creativity, heightened by his consciousness of responding to the unsurpassable self-creativity, the creativity of God.

Autonomy of Man and Religious Dependence in the Philosophy of Henry Duméry

Louis Dupré

If evolution goes from the simple to the more complex the evolutionary process requires a self-sufficiency of its own, hardly reconcilable with the traditional notion of efficient causality, according to which the effect must *on all levels* be inferior to the cause. This difficulty becomes all the more acute when the leap between two successive stages is as considerable as between man and his anthropoid ancestors. Lecomte du Noüy has dealt extensively with this problem: the key to his answer lies in the notion of finalism.

Yet, there is another aspect of the problem which Lecomte du Noüy anticipated, although it was not quite ripe at the time of *Human Destiny;* namely, the relation between divine causality and human freedom. In the meantime this problem has grown up and poses perhaps the most serious challenge to the theistic humanism advocated by Lecomte du Noüy. Few intellectuals today would reject the idea of God because of their acceptance of the evolution theory as such. But many, particularly philosophers, find the idea of a divine causality incompatible with the final product of this evolution, *free man.* To solve this problem the French philosopher Henry Duméry fully accepts the notion of human autonomy in all domains, including religion, and detaches the idea of God from causality.

Evolution is evolution toward freedom. Lecomte du Noüy writes in *Human Destiny*: "Evolution has all the appearances of being a choice, always made in the same ascending direction toward a greater liberty." But once this choice reaches the level of consciousness, it becomes more than a choice among preexisting possibilities; it creates its own possibilities. Such a creativity of modes of existence—which is the very essence of freedom—is incompatible with the

* Part of this address appears in *Faith and Reflection,* a book on the philosophy of Henry Duméry, published by Herder and Herder, 1969, and appears here with their permission.

traditional concept of God as the *fons veritatis et bonitatis*. The question, then, naturally arises whether a fully developed concept of freedom must not exclude any obediential relation to a transcendent being. We all know the radical formulation which Sartre has given of this problem: If God exists man cannot be free—he would only be able to *create* evil since all good preexists him. But even authors who fully accept a relation to the transcendent have always been aware of the opposition between human autonomy and divine omnipotence. Lecomte du Noüy epitomizes this in the paradoxical expression: "God abdicated a portion of his omnipotence when he gave man the liberty of choice."

On the other hand, the existentialist position which accepts freedom itself as an ultimate that allows no further questioning, is arbitrary. For human freedom in its contingency inevitably poses the problem of its own foundation. Duméry admits with Sartre and Merleau-Ponty that the notion of freedom basically conflicts with any preestablished order of values. Yet he adds that autonomy itself requires a transcendent source of energy impelling man to create his own finite determinations.

One of the major tasks of philosophy is to reveal the nature of this ultimate principle, not as it is in itself (for that would only renew the objections of Kant's *Critique*), but as it is present in the creative impulse. The search for the source of creativity is ultimately a search for the One that attracts and impels all cognition, desire, and feeling. According to Plotinus on whose insight Duméry bases his position, the mind is identical with the One insofar as the One enables the mind to posit itself. Yet the One is not the mind insofar as the mind still labors under an insurmountable opposition between subject and object. The One is present in the mind's striving, not in the mind's realizations. To some extent, all religious thinkers are conscious of this tension between the immanence and the transcendence of the One. Lecomte du Noüy expresses it in saying: "It is not the image we create of God which proves God. It is the effort we make to create this image."

Yet, Plotinus is not the only source from which Duméry draws his inspiration. Hardly less important is Blondel's *L'Action*, published in 1893, which shows the logical necessity of posing the religious problem on the basis of the discrepancy between the infinite impulse of action and its limited achievements. "All attempts to bring human action to completion fail, and yet human action cannot but strive to complete itself and to suffice to itself. It must, but it cannot. The feeling of impotence as well as that of the need for an infinite consummation remain incurable."[1] The believer's claim that this completion is achieved in revealed religion is, according to Blondel, at least a logically necessary hypothesis.

For the technical apparatus to express his intuition, Duméry turns to Husserl's phenomenology. He reinterprets the various levels of the phenomenological reduction as stages of the mind's conversion to the One. The *eidetic*, *transcendental*, and *egological* reductions which reduce consciousness to phenomena, related to and produced by the transcendental ego, are essential moments of the *itinerarium mentis ad Deum*. Yet, as Husserl himself admits, they reach no ultimate absolute. Duméry sees the need for a fourth, *henological* reduction which grounds the ego in the absolute One and thereby brings the mind's striving toward unity to rest. The decisive transition between the transcendental and the henological reduction is explained in Duméry's theory of the act-law.

Consciousness alone constitutes meaning and value. Yet the constituting ego is not the empirical self, but a more basic transcendental ego. Duméry refers to it as the act-law. The ego is an act because it creates, produces, and orders. Yet, it does not create at random, independent of rule and law. Although the ego is of necessity autonomous in its determinations, it receives its impulse to determine from an indetermined ultimate, the One. For that reason the ego may rightly be called a law, that is, an act preordained to determine being according to an immanent necessity.

How can the self be a law to itself without being arbitrary? Duméry's answer is that in creating meaning and value, the self reveals its essential thrust toward the One, and that this henological direction excludes all arbitrariness. Undoubtedly, the self may jeopardize the mediating function of his activity by overemphasizing some values to the detriment of all others. Yet this error does not result from the creative act itself, but from man's failure to recognize its relativity; that is, its relatedness to the One, which alone is absolute.

It is important to keep in mind that the ego's creativity does not follow from the awareness of the self, but that it precedes it as its necessary condition. Man constitutes meanings and values before he experiences himself as a constituting ego. Even this awareness does not reveal the act-law directly, but only its expressions. The task of philosophy consists in discovering the ego's original, intentional impulse underneath these conscious expressions and in understanding constituted ideas and values as objectivations of more fundamental attitudes. Philosophy thus brings into reflective focus the most basic activity of the mind from which the empirical self originates.

The distinction between act-law and empirical consciousness is essential. It allows Duméry to eliminate several false problems. One such problem is the opposition between freedom and determinism. As long as consciousness remains restricted to what is empirically accessible, freedom is simply an unexplainable exception in a deterministic world. As the positive sciences (one of which is psychology) draw ever narrower circles around its little enclave, the suspicion grows that some day scientific predictability will cover a field which a lack of knowledge alone still withholds from complete determinism. In Duméry's view, however, scientific determination itself originates in the creative spontaneity of the ego. Instead of suppressing freedom, determinism presupposes it. For rather than merely ratifying a preexisting objective, necessity, freedom for Duméry is the subjective but necessary source of all objective determinism.

Of course, the freedom which produces necessity is by no means arbitrary; it cannot even be identified with the traditional "freedom of indifference." It surpasses the psychological experience of deliberation as well as the unclassifiable phenomena that remain after the scientist has finished his work. If the prescientific concept of unlimited freedom is too simplistic, the positivist notion of an all-comprehensive preexisting determinism is even more so.

> Psychological consciousness knows only those causalities which could originate from its own choices, or fragments of causality of which it does not know the ins and outs, or, finally, the not-further-justifiable shock of factitiousness. Never except by abusive majoration or generalization is it aware of a necessity at once intelligible and impelling. Similarly, the scientist discovers phenomenal connections, experimental concatenations. He supposes determinism wherever nature does not respond *no* to his question, wherever the precision of his calculation allows him to tie together phenomena which must be integrated into an operative whole. But never does he find himself confronted by a fully constituted causalism. Order supposes an organizer. The experimental structures are those of the laboratory, that is, of man and his instruments, not of nature.[2]

If determinism cannot exist without freedom, neither can freedom exist without determination. Freedom and arbitrariness exclude each other. The act-law is essentially order, and order means determination, but the determination is the self's determination, the order is the self's order, not one imposed from without. Even to place God under the denominator of being would make man dependent upon God in the order of determination, and would be incompatible with the autonomy of the act-law. Duméry therefore rejects the traditional notion of participation. Being and the source of being cannot share in the same being. As to the principle of determination, Plotinus' intelligible, it proceeds from the One, but does not participate in the One. Henology and participation exclude each other. "One must choose between those two possibilities. The advantage of the henology is that it simultaneously guarantees the radical productivity of the One and the creativity of the intelligible, the One as 'source-principle' and the intelligible as self-position."[3]

This radical status of the self's creative autonomy raises several questions. The first one is: If God is above all categories, how can we know about him? Any form of "natural theology" seems to be excluded. Even to prove God's "existence" is not possible–much less to discover his attributes. Duméry is not deterred by this negative conclusion, for it forces him to lay all the more stress upon the notion of revelation. Man cannot but make the movement toward God, yet he cannot know God unless God reveals himself. "Because he cannot himself speak our language, all speculative revelation becomes impossible. That is why the Bible declares fruitless all human wisdom which claims to teach us *in toto* or in part what God is."[4]

One may wonder, however, whether Duméry's radically negative theology has not eliminated even the possibility of a revelation. Some critics think that he has removed God so far from man as to make any subsequent contact impossible. Duméry answers this objection in the second edition of *La foi n'est pas un cri*. He admits that God is silent insofar as he transcends all determinations of language. But God can reveal himself insofar as he is the source of our speaking. If the henological reduction can be made, a dialogue with God is possible. Duméry may "reduce" a substantial part of what the ordinary believer considers to be essential to the notion of revelation. But he stresses just as strongly the possibility of and the need for a revelation.

> God is beyond our grasp, our categories, but he appears through them. He is the high point of our aims, he is their soul. Nevertheless, we must increasingly purify our ideas and our plans. For instance, he cannot be called *personal*—personal as we are; but he is more. It is possible, it is normal to address ourselves to him as to a person, provided we preserve the mystery.[5]

Definitely eliminated in Duméry's thought is a revelation of God through nature. Nature may help man find his way to God, but it can never teach him anything *about* God. Nature has no voice of its own–all revelation is essentially human, for man alone can give meaning and expression. So if God is to speak at all, he must do it through man. Man alone is the image of God.

But this brings us to a second question. If man alone has the power of speech and if God is above all human categories, how can any revelation be said to be *God's* word? Revelation seems to be an expression of man, in both its content and its form. In constituting the sacred, the human subject creates mediating schemas and categories in order to attain the Absolute. As expressions of a human experience, these schemas and categories are obviously human. So, then, what entitles the religious man to read God's word in them? If they are merely intermediate stages in the mind's ascent to the One, they are at best relative expressions. How, then, can the believer ascribe a permanent, absolute meaning to the words of the Revelation? How can they have a permanent, absolute value? If the One is above all intelligible determination, why should the New Testament be anything more than the revelation to the Absolute of one particular cultural group expressed in accordance with their specific needs and aspirations? Duméry fully admits the relativity of all religious language, including that of the New Testament. Every language bears the imprint of the civilization which it expresses, and that civilization is, by its very nature, relative. But he denies that this relativity eliminates the absolute element. The transcendent nature of God does not exclude an objective revelation—nor does the subjective acceptance and expression of this revelation, the so-called projection, eliminate its objective character.

> Consciousness does not project anything *upon* the object, it does not cover it with something that does not belong to it. I use the term projective consciousness as opposed to reflective consciousness. I mean that consciousness itself is spontaneously projective: it projects, not something of itself upon something other than itself, but the meanings which it intends on a diversity of expressive levels. Its act is intentional: it is directed toward the object and attains it; but it cannot intend it without expressing it at the same time in a spectrum of various representations.[6]

Subjective structures do not change the object: they are the indispensable means by which the subject attains objective essences on various levels of consciousness. They are the prism in which the object itself is refracted. That the New Testament projects the religious consciousness of a particular community, by no means implies that it has deformed the objective character of the *fact* Jesus. Indeed, without such a projection there would be no religious history. For what transforms these particular facts into *religious* facts if not their mode of acceptance? For meaning, history must depend on meaning-giving subjects. This is particularly true in the case of sacred history.

The events of Jesus' life can have a religious meaning only to a religious subject. To *understand* the religious meaning of Christ, it is not sufficient to register the historical facts of his existence. "The religious reality objectively contains the meaning which the believer recognizes in it; yet, this meaning is perceived only when the believer discerns this reality as religious. In technical terms: the religious object exists, but we still must 'constitute' it as religious."[7] The most essential characteristic of the religious object is that it must be received in a religious way; that is, that it must be *given* a religious meaning.

Far from being a deviation from the original message, the interpretation of faith is a necessary factor for the correct transmission of this message. For the message itself refers to faith and this cannot be transmitted by merely factual report. It requires a personal commitment on the part of the reporter. If the Evangelists had not told their story in a spirit of faith, their writings would have been no more "sacred" than the brief reports on Christ in Tacitus and Pliny. Historical reliability requires that the events be rendered as truthfully as possible—not that the narrator abstain from all religious interpretation, for this interpretation is an essential part of the events *insofar as they are religious.*

The religious interpretation also justifies the selection which the sacred narrators apply to their historical material, and which must appear quite arbitrary to the nonbeliever. Even many believers who imagine that it suffices to "get the facts" in order to have faith are shocked by this selection and prefer to ignore it. But every history requires *some* selection and since a "religious fact" can be recognized only in a religious vision, a selection of facts on the basis of their religious acceptance becomes imperative. Duméry's position supports the conclusion of contemporary biblical scholarship, that the sacred writers found their inspiration within a religious tradition and must be read in a religious tradition. The tradition is the faith of the witnessing community which is able to bestow the religious meaning upon historical events.

Yet the necessity of a subjective acceptance in faith, of a tradition, does not reduce religion to a merely subjective experience. Christianity has always strongly emphasized the historical character of its foundation. The object of Christian faith can be seen only through the eyes of faith, yet, faith itself requires that its object be *historical.* The religious vision of faith is empty without the historical facts which it illuminates. An attack upon the historicity of the basic events of Christianity, Jesus' death and Resurrection, is, therefore, an attack upon Christianity itself. No doubt, faith in the Resurrection goes far beyond the historical apparitions and the discovery of the empty tomb: it demands that death and Resurrection be accepted as essential stages of the revelation of the Son of God and of the redemption of man. Yet, the act of faith itself needs the foundation of historical facts. Only a dialectic of seeing and believing could ever lead to the Pentecostal experience.

> The apparitions of Christ are proofs because they do not merely present the risen Christ to seers, but to seers who are also believers and whose belief brings them to the proper perspective, to an order of truth in which it would be contradictory to claim that Jesus is the Christ, the blessed of God, without having broken the bonds of death.[8]

The apostles believed because they saw and they saw because they believed.

Another point that must be emphasized against any subjectivist interpretation of man's theophanic activity in Christianity is that the original and most basic meaning of Jesus' acts was given by Jesus himself. This meaning was reconstituted and developed by the primitive Christian community. The result was transmitted to later generations who, in turn, reconstituted these objective data into religious experiences. Since the meaning-giving activity is obviously conditioned by the personal characteristics and the cultural level of the interpreters, the religious interpretation of "the fact Jesus" varies from age to age. Yet, all these variations do not basically deviate from the original meaning given by Jesus himself. Against the position of form criticism, Duméry maintains that the collective consciousness may enrich an idea, but that the idea itself can originate only in a personal consciousness. What the Christian community sees in Christ was, at least implicitly, immanent in Jesus' experience. To recognize Jesus as the Lord is an act of faith that no historical "facts" can "substantiate." Still, the Christian's belief cannot be without objective support. For why is it that the believer's faith centers exclusively upon Christ, while he rejects all other "theophanies"? Whatever the basis of such a religious discrimination may be, it cannot be purely subjective. Duméry remarks:

> The religious consciousness defies only those beings that display for its eyes a presence indicating divine authority. Every theophany is a value judgment, the value of which is proportionate to the spiritual requirements of him who formulates it. How could we deny that in the case of Christ and his disciples the theophanic judgment passed on Jesus was inspired by his own attitude through which it attained a particularly pure conception of the nature of God.[9]

Some objective religious meaning must obviously be transmitted. Since such a meaning is not immanent in the historical events as such, it must be placed in them by a religious interpretation of the events. In Christianity, this interpretation started with Jesus and his first disciples.

NOTES

1. Maurice Blondel, *L'Action* (Paris, 1893), p. 321.
2. Henry Duméry, *Philosophie de la Religion*, I (Paris, 1957), 60.
3. Henl van Luyk, S. J., *La Philosophie du fait Chrétien* (Paris, 1965), p. 78.
4. Henry Duméry, *La foi n'est pas un cri* (Paris, 1959), p. 216.
5. *Ibid.*, 2nd edition, p. 225.
6. *Ibid.*, p. 244–245.
7. *Ibid.*, p. 258.
8. *Ibid.*, p. 83.
9. *Ibid.*, p. 74.

Commentary

Thomas Langan

Professor Dupré quotes Duméry as declaring that "apparitions of Christ are proofs because they do not merely present the risen Christ to seers, but to seers who are also believers and whose belief brings them to the proper perspective . . . in which it would be contradictory to claim that Jesus is the Christ . . . without having broken the bonds of death." Dupré paraphrases this: "The apostles believed because they saw and they saw because they believed." But what Duméry said in this quotation was, "They saw what they saw because they believed." And that is not the same thing.

This difficulty arises because in Dupré's exposé of Duméry there appear two distinct positions. Early in the paper, we glimpse an emanationist Duméry; in the second half, a phenomenological Duméry. The first position raises the problems of an emanationist fideism. The second, the classic problems of interpreting form and interpreted matter one associates with the Husserlian position.

I shall take a minute to discuss each of these problems, and then I shall ask Professor Dupré whether the effort to bring these positions together and solve the problems does not plop one right into the lap of Hegel.

Early in the paper the transcendental ego is described as creative, producing, ordering, according to an act-law—"an act preordained to determine Being according to an immanent necessity," and which of its very nature is an essential thrust toward the One. Such a transcendental ego, issuing totally from the ultimate unity, would appear to create its own object. Error in this scheme of things would not, as in the phenomenological, be a failure to harmonize an interpretation with a core of to-be-interpreted, given matter. Rather, error would arise from interference by a negative finite ego with the necessary process of total revelation through the transcendental ego, which egoistic interference temporarily restrains the absolute from proceeding in its course of full self-revelation.

The objections one needs to raise to that position add up to showing that the struggle to know, while resulting necessarily in distortions, incomplete reports, unfruitful relatings, is still a positive activity and not just an interference with a revelation through growing consciousness which proceeds necessarily and of itself. Moreover, if all efforts of believers to understand are due wholly to divine impulse, so that all sincere creeds are in a sense revelations of God, one wonders by what criteria one would distinguish the truer from the less adequate position.

The phenomenological position which emerges in the latter pages of Dupré's paper is not like that at all. There the object, while the result of an interpretative-constituitive act of the transcendental ego, is not entirely the product of it, for the matter (in Dupré's example, the historical facts) is given, so that the interpretation must be in some way appropriate to the matter.

For classic positions there are classic problems. I shall state the main difficulty in the terms suggested by Dupré. The constituted religious object—in this case the conception of Christ—would be empty, we are rightly told, without the historical facts. But then the interpretations of the believer's consciousness must somehow respect these given facts. How, then, explain that these facts get themselves presented in such a way that a certain range of interpretations, developed in the course of history, is suitable, while others would presumably have to be ruled out as irreconcilable with the facts.

The stronger emanationist position avoids the difficulty by affirming that the transcendental ego creates its object entirely—matter as well as form. But it pays the price, we have just noted, of rendering the struggle to know illusory.

Does Dupré mean to suggest a way out of the Husserlian interpretive form-interpreted matter paradox when he writes, at the end of the paper, "Only a *dialectic* of seeing and believing could ever lead to the Pentecostal experience"? But no dialectical explanation avoids the chicken-and-the-egg problem of facts guiding the deployment of relevant interpretative categories, without which those facts themselves have no sense.

If Duméry would avoid the problems of emanationism by suggesting that the one transcendent source is present in the world *first* as alienated nature and *then* as creative impulsion of the transcendental ego which "still labors under an insurmountable opposition between subject and object," then how, Mr. Dupré, does this position differ from Hegel's suggestion that spirit is alienated in nature and present in finite consciousness as the impulse to recuperate the sense that is in nature? For Hegel too, God is revealed through *our* thought of nature and ourselves.

Lecomte du Noüy as Biophysicist

Ralph W. G. Wyckoff

Scientists will think of Lecomte du Noüy as one of the pioneers of biophysics; nevertheless in considering his work, we must realize that the biophysics envisaged by him—and others of us who a generation ago thought of ourselves as biophysicists—was rather different in scope from what it has become in the intervening years. Only by taking account of this will we have a proper sense of what Lecomte du Noüy was seeking to accomplish through his scientific effort. When he was actively at work in the laboratory forty years ago, biochemistry was starting to emerge from physiological chemistry as a broader study of the chemical composition of all forms of living matter. It was just beginning to isolate pure enzymes and other complex substances from cells and tissues, to analyze them chemically and to establish their roles in the life of these cells. In order to understand any natural system, inanimate or animate, it is necessary to know how it works as well as its composition; and in those days it was imagined that a biophysics, as yet unformulated, would reveal the mechanics of living matter.

The physics of inanimate matter was developed by making quantitative measurements of its behavior under a wide range of conditions. Knowledge was gained by devising experiments that could be repeated at will on carefully isolated and simplified natural systems. No corresponding freedom of choice is possible when experimenting with living organisms. They cannot be broken down into simpler parts without destroying the life that is their distinguishing characteristic; and significant repeatable experiments are rare simply because little of importance can be done to a living system without changing it irreversibly. This is why, as Lecomte du Noüy fully realized, biophysics cannot be developed as an extension of the physics that has given us the laws of inanimate matter. The laws of biophysics surely will not conflict with those of physics, but they must grow out of a different kind of experimentation, and they will have their own unique forms. Lecomte du Noüy was groping, as we still are, towards a clearer picture of what we must know in order to understand the basic principles of life.

This search for basic principles, becoming more clearly evident with time, can be traced throughout his life. The same thing is to be seen in the life of every creative person; but in the scientist it is particularly direct because he, even more than the artist, acutely feels the need to rationalize what he is doing. Scientists like Lecomte du Noüy are explorers; through their work they seek things that have a deep and direct connection with life's meaning. We may be sure that after devoting himself for many years to other things, he turned to the scientific laboratory because he believed that what he could find there would clarify his understanding of this meaning. Such was the motivation of many who, at the turn of the century, looked on science as *the* way to satisfy an essentially mystical search for the significance of their lives. The pursuit of meaning through a preoccupation with nature is as old as the complementary religious search through cultivation of the inner life. Many of the founders of modern science–for example, Newton, Kepler, and Copernicus–sought God through the investigation of nature; and this intimate interplay of the life of the senses and of the spirit was not lost with the realization that God is not to be approached as the day-by-day manipulator of the universe of matter. We have only to examine the lives of such more modern scientists as Faraday, Pasteur, or Planck to appreciate how inwardly motivated may be a life of science.

At the same time there have always been sincere materialists who find the totality of their human experience satisfied by the hypothesis that nothing exists apart from the world of matter perceptible to our human senses, that what we call the inner life of the spirit is merely a by-product of the physical and chemical activities of our material bodies. Such persons have gained much support for their outlook from the general acceptance of evolution and, more recently, through biochemistry's success in explaining mechanisms of the living process. We who were born toward the end of the last century grew up under the influence of both this burgeoning science which seemed to offer the answer to every question to which it seriously addressed itself, and the emphatic presentation of its materialistic interpretation. Under the impact of the success of science, we turned to it as the way to understand life and we came to expect from it far more than it can give. Science was then curing disease after disease and discovering the laws of nature which in the hands of engineers were suddenly giving man a wealth and ease of life which even our fathers could not have imagined. We were overwhelmed by the promise of this success and felt life's greatest privilege was the opportunity to participate in scientific endeavor. Many accepted without question the materialism that flourished in this atmosphere; others who by temperament and individual experience were less easily convinced have had the lifelong problem of integrating into a rational whole their culturally inherited attitudes and a personal experience which included a steadily improved picture of the natural world and how it works. Lecomte du Noüy was a man of that generation and we can see in both his scientific work and in his later more philosophic writings steps in his own integration.

As Mrs. Lecomte du Noüy has pointed out in her biography of him, he was strongly influenced by science long before becoming a practicing scientist himself. Time spent in the laboratories of the Curies and of Sir William Ramsay is testimony of this. Earlier in the last century Claude Bernard was one of the first to emphasize the need for biology to become quantitative before it could be considered as anything more than a description of nature; and a drive towards accurate measurement was already being undertaken, especially by the English school of physiologists. The proposal of Carrel during the First World War that Lecomte du Noüy should seek a quantitative description of the rate at which wounds heal had a strong appeal for him. Undoubtedly, it was his success in dealing with this problem, half practical and half biophysical in the sense of using physical modes of thought to attack a matter of biological import, which led him to devote the rest of his life to what was then looked on as the beginnings of biophysics.

Lecomte du Noüy's scientific career was the natural outgrowth of these initial studies of cicatrization. A preoccupation with the properties of blood serum, which is of course the medium within which healing takes place, pervades his subsequent investigations and plays an obvious role in the gradual evolution of his estimate of the place of his science within his total experience of life. I recall his saying, when our acquaintance was still young, that serum is the conveyer of life: it nourishes all the tissues, it washes them free of the waste products of their activities, and it supplies them with protection against the infectious diseases that otherwise would destroy them and the organism of which they form part. It is the internal environment which supports life in ourselves and in the cells of which we are composed. To Lecomte du Noüy there could be no more fundamental problem associated with life than the study of this serum.

The first problem of serum with which he concerned himself was its surface tension, and Mrs. Lecomte du Noüy tells us that its investigation was already underway in the hospital in Compiègne. It began with the invention of an instrument for better measurement. Modern science has grown through the invention of instruments that extend the range and scope of our sensory perceptions of nature and enable us to measure in a quantitative fashion what is taking place. Without this quantitative information few of the laws of nature could have been discovered, and without these extensions in scope, our picture of the universe would have remained as restricted and faulty as in ancient times. Pioneers in science have therefore been intimately concerned with the building and application of new instruments; in this respect Lecomte du Noüy was very much a pioneer. He began his studies of surface tension by developing a better instrument for its measurement. His first results with it were described in the *Journal of Experimental Medicine*, from the Rockefeller Institute for Medical Research, soon after he established his laboratory there at the close of the World War I. These results, dealing with the reversible decrease in the surface tension of serum when it stands, were the starting point for a long series of investigations. They demonstrated how monomolecular films develop on the surface of protein and other colloidal solutions, and how the dimensions of these films and the molecular weights of the macromolecules forming them can be ascertained.

It would be impossible in the time available to us even to suggest the broad scope of Lecomte du Noüy's experiments dealing with this subject over the better part of a decade, but they must be considered an epoch-making example of the new and informative results to be obtained when a biological system is subjected to a thorough physicochemical investigation. It should be noted that he was not content to rest with the examination of normal serum; he proceeded through studies of immune serums to demonstrate the direct application these biophysical methods can have to pressing problems of pathology and medicine. Nowadays we take this for granted, but when he was carrying out his experiments, most biologists and medical men were yet to be convinced that such methods had much to contribute to the understanding of vital phenomena.

The practically instantaneous measurements of surface tension that could be made with the du Noüy tensiometer gave it a field of application reaching beyond the biological fluids that were his primary interest. It has, for instance, been extensively employed in the study of lubricating oils, and hundreds of these instruments have been manufactured and sold. He never lost interest in its applications, but his own research soon moved to other properties of serum. The first to gain his attention was viscosity, and the effect of temperature and composition on it. For such measurements, too, a new instrument was needed, and soon after coming to the Rockefeller Institute he designed and built one. He investigated in detail the profound and irreversible change that occurs in all serums when heated to temperatures between 55 and 60°C, and he interpreted this change as a consequence of first steps in the breakup, or denaturation, of the protein molecules of serum. According to his ideas, chemical bonds break within the molecules at this critical temperature, thus permitting the introduction of more water to yield a higher viscosity. He was also able to demonstrate, with his highly sensitive and accurate instrument, the course of the chemical reaction that takes place when an immune serum interacts with the antigen against which an animal has developed its protective antibodies. As far as I know, this and his studies of surface tension were the first successful quantitative physicochemical investigations of the basic immune reaction that is our fundamental defense against infectious disease.

As time went on and he returned to Paris to establish a laboratory at the Pasteur Institute, Lecomte du Noüy undertook the study of still other properties that varied with temperatures and with the composition of serum. He found that better measurements of acidity were required, and once more built an instrument that would give more accurate values for serum. Still other sensitive and vitally important properties of serums are such optical characteristics as index of refraction, light scattering and absorption—from the infrared through the ultraviolet. In the Paris laboratory Mrs. Lecomte du Noüy actively aided her husband in the development of this aspect of his work. Their collaboration there also included the introduction of the techniques of tissue culture brought from Carrel's laboratory in New York. It would be inappropriate for me to here try to tell you about the growth of this biophysical laboratory at the Pasteur Institute; we can only regret that after so auspicious a beginning it should have become a casualty of the Second World War.

As fate would have it, this was the last chapter in Lecomte du Noüy's active scientific career. After escaping from France in the midst of the war and establishing a home in Altadena, they began to build a laboratory in which he planned to continue his interrupted investigations. Final illness overtook him before this could become productive but he had installed the instrument-making shop which always was at the heart of his scientific efforts. Mrs. Lecomte du Noüy gave me this shop to use in connection with my research since, like Lecomte du Noüy, I must build the instruments I use. This shop was my office and private laboratory at the National Institutes of Health and it has become the focal point of the larger laboratory I have organized on retiring to the University of Arizona. In that important sense his equipment is continuing to further the outlook on science and its goals which we so deeply shared.

As I have already remarked, biophysics is currently developing in directions which are different from those we were pursuing a generation ago. This is due primarily to the invention of new techniques that have allowed scientists to devote themselves to questions which could not earlier be answered. Physical knowledge of inanimate matter has come through the study, first of matter in the gross and then of its microcomponents, together with light, x-rays and the other forms of radiation with which it interacts. A similar development may be expected for biophysics. Lecomte du Noüy investigated gross properties, and this aspect of biophysics is far from being exhausted. Quantitative studies of the interaction of living organisms with radiation, like those begun in the laboratory at the Pasteur, have been greatly stimulated more recently by the growth of atomic physics. Nevertheless biophysics today is concentrated on the so-called molecular biology which deals with microstructure. It has been made possible by the development of new instruments to isolate and investigate many of the characteristic macromolecular substances of living cells whose coordinated physicochemical reactions are essential to life. This is describing so much of the mechanics of the living process at the molecular level that for the moment it has drawn to itself scientific attention which would have gone into the expansion of other aspects of biophysics. Such knowledge will of course lead to a better understanding of the functioning of the cells and serums to which Lecomte du Noüy devoted himself.

This is not the place to outline molecular biophysics or to suggest what its future may be. It now centers around the nucleic acids to such a degree that any discussion of biophysics must at least mention them. In what real sense are they (as materialists assert) the key to an understanding of life? As we observe them in nature they play a role analogous to that of the coded tapes which guide the functioning of a modern automatic machine or factory; the complexities of their chemical structure appear as a code that corresponds to the pattern according to which the cell, thought of as a chemical factory, operates. The mechanism for accomplishing this is still obscure, but one cannot reasonably doubt that there is a detailed parallel between the molecular structure of the nucleic acids of a cell and its genetically inherited capabilities. The cell is more than a machine in that it is self-perpetuating, and undoubtedly this added capability finds its reflection in the observable structure of its nucleic acids. Many present-day biologists imagine that this nucleic acid code is leading to a full explanation of life in atomic and molecular terms; we shall not debate this. However, it should be noted that codes may specify how a machine or cell works, but not what it is; and we must emphasize the wide, unbridgeable gap that exists between such a belief and what is actually observed.

The rest of my time could profitably be used in an extended analysis of the problems briefly outlined above and in a more detailed account of that aspect of biophysics which Lecomte du Noüy was influential in initiating. But he has published nearly 200 scientific papers and books, and there are other consequences of his devotion to science which I prefer to discuss. In doing this, many of my own ways of seeing things will be mixed with a paraphrase of his ideas; but knowing him as I did, I feel sure that what I shall be saying will not be counter to what he himself might have said.

Modern science is the fruit of a consuming effort by a relatively few men to penetrate the mystery of the universe and ascertain the real place of man within it. Thus it represents a quest that has been as much religious as philosophic. In recent times this primary search for understanding has been supplemented by other objectives. With scientific knowledge has come the ability not only to understand but to manipulate nature for material and often selfish ends. As a result, science is rapidly becoming a dominant factor in the conduct of human affairs, and like all other reservoirs of power, it can be used equally for good or for evil. Inevitably, the new element thus introduced into practical affairs is attracting to the life of science persons differently motivated from those who created it. As a result, science has become one among many professions. The material wealth that flows from new natural knowledge has indeed accelerated the growth of science, but we need to recognize that it has at the same time transformed the scientific community and given it many controlling objectives different from those that created it.

We scientists are reacting very differently to this growth in the content and social importance of natural science, Many—probably most—are finding it satisfying and have been fortified in that sense of its overwhelming importance which drew us to it in our youth. Those whose outlook on life is materialistic feel this way for it expresses their underlying assumption that science is unlimited in scope and able ultimately to interpret all aspects of human experience. Others who have realized the incompleteness of science find two roads open. If they have adopted a scientific career because of its newly acquired professional status (or more idealistically, because of its potential benefits to society) they are naturally drawn by the many social questions that result from the power conferred by scientific knowledge. It is obvious that our society urgently needs such people, but in gauging their activities it is important to remember that sound training in science is perhaps the poorest imaginable preparation for a life of practical politics. Politics involves the art of compromise, and a truly great politician must be a man of high and realistic principles who is skilled in this art and can practice it without tarnishing his standards. In the pursuit of science there is no room for compromise. Its observations and explanations are either true or false. A good scientist has little preparation, and perhaps little aptitude, for a life of compromise. Lecomte du Noüy's appreciation of the incompleteness of science did not lead him into social and political activity. He took the other road, seeking a better understanding of those aspects of life which for him were not covered by science. His accumulating experience was not satisfied by the philosophically simple outlook on reality which is provided by an all-inclusive materialism. Throughout his later life and work he sought to express this dissatisfaction which is at the root of the theory of telefinalism as stated in *Human Destiny*.

Modern science must be recognized as one of the great creative outbursts of the human spirit. In common with great art, it calls for the most disciplined observation of nature. It has also stimulated the equally disciplined intellectual activity and imagination required to interpret correctly its quantitative observations. The pursuit of science is indeed a reciprocal thing: as individual scientists we can add in some measure to an understanding of the processes of nature and in so doing further the well-being of mankind, but we are ourselves molded through this activity. What we become as persons depends above all else on our goals in life. If our sights are set high enough, the training and self-discipline involved in a fruitful life of science can bestow a capacity to see ever more deeply into the still unfathomed mystery of existence and the meaning of human life. Lecomte du Noüy's vision of science as one of man's great adventures made his work the path of steady growth as a man and broadened his understanding of life. This is readily apparent in his later writings; and as we all know, *Human Destiny* expresses with extraordinary clarity his point of final attainment. Nevertheless the scientific approach is incomplete. The tumultuous success of science was for his generation, as it still is, an ever-present danger as well as a supreme opportunity. It has tended to monopolize our thoughts to such a degree that other aspects of life usually do not receive the attention and development they deserve.

Science is insufficient because it necessarily excludes those values we all experience as an essential ingredient of life. It must now be clear to all thinking persons that our mental lives, like our physical bodies, have evolved from very simple beginnings through sensory adaptation to the environment furnished by this earth. Life for all animals is a series of decisions which at the primitive level are no more than instinctive reactions to sensory contacts. We can observe in ourselves and the higher animals how the gradual development of consciousness transfers these reactions and decisions progressively from the instinctive to the willfully determined. We also see their dual character: some are based on the material realities, the truth, of a situation; others on desirability.

In the conscious life of man the rational intellect represents our way of coping with questions of truth, while our criteria of desirability have evolved out of purely biological urges to include a more or less disciplined sense of esthetic and spiritual values. The history of human culture is the story of the gradual development of these two components of that inner life which the emergence of consciousness has made possible. In man today, natural science has the task of defining, criticizing, and expanding our inherited picture of reality within the universe of matter; it does so by applying the criterion of truth to sensory experience and to the intellectual operations by which such experience is interpreted. Other qualities of mind, cultivated through other disciplines, are required to deal with that world of values which complements and should take precedence over the developed human being's understanding of nature. The faculty for deciding about values which has arisen in the conscious mind of man is as responsive to discipline and as rational as the intellect that decides about truth; but the rules under which it operates are entirely different. The decisions of science are unequivocal: as I have said, a proposition describing nature is either true or false. On the other hand, there are gradations in decisions about values: innumerable steps lie between the ugly and an unattainable beauty, between folly and wisdom, or between the man who practices evil for its own sake and the saint. The faculty that characterizes a great artist, a wise or a holy man is no less rigorously developed and disciplined than that of a dedicated scientist, but it works in a different way. It expresses another aspect of experience and requires another schooling for its growth and fullness of expression. That is why the current worship of science is a danger both to the practicing scientist and to the continuing growth of mankind. Science is important and must continue; but it should be seen in its proper, limited perspective. By pursuing unrestrainedly the pleasures, intellectual as well as material, that follow from the cultivation of natural science, we shall lack the time, even though we may have the inclination, to cultivate our higher values. It behooves us to remember that the greatest expressions of the human spirit have been and always will be those that involve both truth and value. This is the point of view at the root of Lecomte du Noüy's sense of the importance, but incompleteness, of science.

Earlier cultures have confusedly intermingled their pictures of the outer reality and their worlds of values; one of the great contributions of modern science is the gradual removal of this confusion. The lives of all men have been and still are surrounded and penetrated by the mysterious unknown; and most men have personalized this unknown as their gods. Our forebears attributed to their God all that was unknown both within themselves and in the world of matter. It seemed natural that He should be burdened with such tasks as keeping the universe in operation and populating the earth with life; for generations men sought to prove His existence through His engagement in these tasks. By discovering that everything in the material universe follows the dictates of immutable law, modern science has revealed this as a misdirected quest; but we should never forget that in spite of increasing knowledge we remain as enveloped as ever in ultimate mystery. As the definitions we make of energy, time, and life illustrate, the mysterious permeates our basic concepts about nature and should be ever present in the minds of those engaged in scientific research. Men who seek consciously to cultivate their inner, spiritual lives find themselves equally confronted by it. It is one of the great merits of *Human Destiny* that it conveys in so lively a fashion the way this sense of mystery animates man's greatest endeavors.

Those who recognize the dual worlds of sense and spirit in which we all live must wonder about the realities of the seemingly different experience that underlies each. We scientists who readily recognize our sensory contacts with nature as *the* direct experience upon which all knowledge of the material world is based cannot escape asking ourselves to what degree the spiritual is also a consequence of the sensual. Evidently many but not all the primary data of esthetic experience are sensory, and at least some of our spiritual outlook develops out of our contacts with other living beings. However, there have always been men of high and disciplined spirituality who have insisted on their direct experience of something greater than themselves. Their conviction of the reality of a spiritual life apart from and transcending the life of the body may not lend itself to scientific proof or disproof; nevertheless the remarkable transformation in personality seen in those who rightfully lay claim to such experience is as objective a fact as tomorrow's sunrise. Millions of lesser men draw strength from the fleeting contacts they can make through prayer and meditation with this aspect of the inner life. Materialistically oriented scientists can, if they wish, question the origins of this source of inner strength and inspiration, but its reality cannot be doubted. Contact with it grows with man's conscious effort and, as with everything else in life, only those with experience are in a position to have valid judgments.

In the past, men convinced of the reality of an immaterial world of the spirit have sought scientifically demonstrable contacts between it and the world of matter. Two generations ago, a number of distinguished scientists devoted much time and effort to the careful investigation of psychic phenomena that to them suggested a life of spirit apart from the human body. Many of the facts they uncovered carried conviction to them, but are either forgotten or discounted by most present-day scientists. Lecomte du Noüy did not rule out the possibilty that evidence for the immaterial might be found in material events; and this led him to detailed calculations of whether or not life could have originated purely through chance. Though knowledge since gained of the simplest forms of life and of the conditions that may have prevailed on earth in the long distant past has shifted the weight that can be given to arguments such as those he adduced, they have helped many to see more clearly than before where the real problem lies. I doubt if our fundamental understanding of the meaning of life would be greatly altered were we tomorrow to discover how to produce in the laboratory a chain of events that would lead to living forms. Once having fully absorbed the fact that everything in the world of matter obeys its laws, it seems self-evident that the question of a reality apart from sense-perceived matter must be answered by evidence of a different sort. Clearly direct spiritual experience is the proper test of its own reality.

In the course of evolution the brain and nervous system of animals have gradually developed till in man there has emerged a creature whose life is centered in the inner world of mind. With his mind man is learning to picture and comprehend the universe of matter and energy and with it, too, he is enabled to enter a world of beauty, goodness, and wisdom that is a far cry from the simple biological urges which were its starting point. An individual's experience may be such as to satisfy him that this inner life is indeed man-made and a by-product of the biochemical activities of his body. Or he may sense it as somehow illumined by a superhuman world of the spirit which he is at times privileged to glimpse. He will then, with Lecomte du Noüy, see a purposiveness in evolution; it will appear as the process gradually producing an instrument through which spirit can more and more deeply pervade and animate the material universe. To serve as that instrument then becomes man's supreme opportunity and destiny. Scientific knowledge alone does not lead to a choice between these two attitudes; the one adopted will be influenced by the disciplined effort each person makes to develop both aspects of his mental life. For those who have in this way experienced—in even the most fragmentary fashion—a sense of something greater than themselves, the world of spirit takes on a convincing reality that determines the meaning one attaches to life. What Lecomte du Noüy wrote must be viewed from this standpoint.

But whether or not we share in this latter (which to me is a more satisfying attitude towards life) we must all realize that the faculties which have evolved to give man a conscious inner life have made him inescapably a partner in directing the future of all earthly life. Now that we are able to control in mounting degrees the processes of nature, and especially to modify ourselves both physically and mentally, the course of evolution is radically altered. We have thus acquired a measure of responsibility for the further development of mankind and of all earthly life which we cannot evade. In a broadened sense, what we and our children become will from now on depend on what we ourselves strive to become, and our individual goals in life will determine in quite a new way the future of this planet. The men of the Enlightenment, in developing their new ideas of progress, imagined this goal to be happiness, and their emphasis on material well-being continues to dominate the activities of our societies today. Such well-being is not to be scorned, but as *the* goal it is unworthy of man. The characteristically human faculties that have arisen through natural evolution, resulting in a consciousness that creates the inner life, reside in the individual, to be further developed through his own effort and self-discipline. We simply do not have the automatic perfectibility which our present society, following its eighteenth century mentors, imagines will be the response to suitable legislation. Because man cannot avoid determining the further course of earthly evolution, it is imperative that we realize now and quickly that the destiny offered us will be achieved only to the degree that every individual develops his own potentialities. The future of mankind will be determined by what each of us makes of himself. This individual responsibility cannot be shirked, nor can it be delegated to others; it remains the same for each succeeding generation. The way we formulate it, however, shifts as new knowledge is acquired about the material world and about ourselves. Each generation must restate the unchanging fundamental realities in terms of the changed outlook brought about by a greater knowledge, which now is increasingly the fruit of biophysics. Lecomte du Noüy did this in a conspicuously clear way for his generation; his success should give heart to the effort we must each make to keep these realities unclouded.

Biological Time and Aging

Gerard Milhaud

Mankind has always wanted to have two elusive things: wealth and youth. This double dream was expressed by the alchemist in his search for the philosophical stone which he thought could change base metals into gold, and in the hoped-for discovery of the elixir of life, giving eternal youth.

Drugs for rejuvenation are still unknown, despite apparently successful tests made by Brown Sequard with endocrine transplants, by Bogomoletz with reticuloendothelial serum, and by Niehans with fresh cell injections. In a field where psychotherapy plays a major role, therapeutic effects have to be precisely assessed in order to exclude any foreign interference or placebo effect. The value of the mentioned treatments has never been established. Studies of the aging process are among the most important subjects in current medical research, in a world of more and more extended life expectancy. The importance of cardiovascular diseases and of cancers is well recognized. But if new therapeutic agents could cure all cardiovascular disorders and all cancers, the length of life would increase on the average by only seven years, making the study and prevention of aging processes more urgent.

Pierre Lecomte du Noüy made a major contribution to the problem under discussion in introducing the concept of biological time in his book *Time and Life*, published thirty-one years ago. Since then, the problem has not been resolved to any appreciable extent, whereas mankind has experienced the impact of atomic energy, the use of computers, and the conquest of space.

In 1915, Lieutenant Lecomte du Noüy was asked by Dr. Alexis Carrel to establish the law of woundhealing in order to determine the relative value of the treatments in use, the effects of various antiseptics, and the influence of factors accelerating or delaying woundhealing. This process is simple only at first glance: any skin lesion will initiate the woundhealing process; it will stop with the repair. Cells proliferate rapidly, divide, repair the lesion, and then resume their normal activity. The arrest of cell division must be due to contact inhibition: when a cell comes close to another cell, a "signal for information" is received, which leads to the slowing down of mitotic activity. Cancer cells apparently are not subject to this regulation; they are not sensitive to contact inhibition. They continue to divide and can invade neighboring tissues.

But we shall return to the normal process of woundhealing and see how Lecomte du Noüy succeeded in making a basic discovery while working on a problem which required immediate application. The wounds had to be kept sterile during several weeks or months. The area of the wound was measured at regular time intervals with the intention of finding the mathematical relationship linking wound area to time (Figures 1 and 2). After some trials, he established the following empirical formula:

$$S' = S\,[1 - k\,(t + \sqrt{T})] \qquad (1)$$

where

S = wound width at time T

T = age of the wound

k = proportionality factor

t = number of days elapsing between the measurement of the surfaces S and S'

This formula enables one to calculate the area of the wound at any time. It has no heuristic value in itself and is equivalent to the graphic representation of the healing process, which follows a simple law even though it is the result of many factors governed by more complex laws. The investigation of the proportionality factor k led Lecomte du Noüy to discover an essential biological process. K is constant for a given wound but varies from one wound to another and from one man to another. Lecomte du Noüy observed elevated values of k for: (1) smaller wounds as compared to bigger ones, implying that smaller wounds heal more quickly than do larger ones; and (2) younger men as compared to older ones. K is a function of the age and of the area of the wound.

Lecomte du Noüy described the family of curves linking k to the wound area for the age interval of twenty to forty years. It therefore became possible to assess the age of an individual knowing either the rate at which healing occurred or the values of S and k. Aging

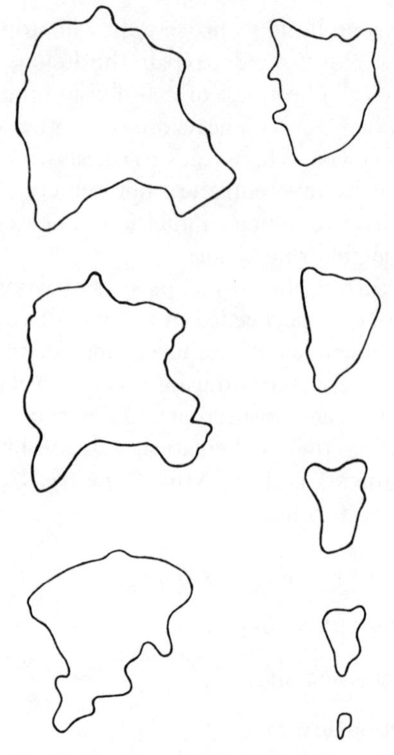

FIGURE 1. Areas of wound measured every four days. In Lecomte du Noüy, *Le temps et la vie* (Paris: Gallimard, 1936), p. 108).

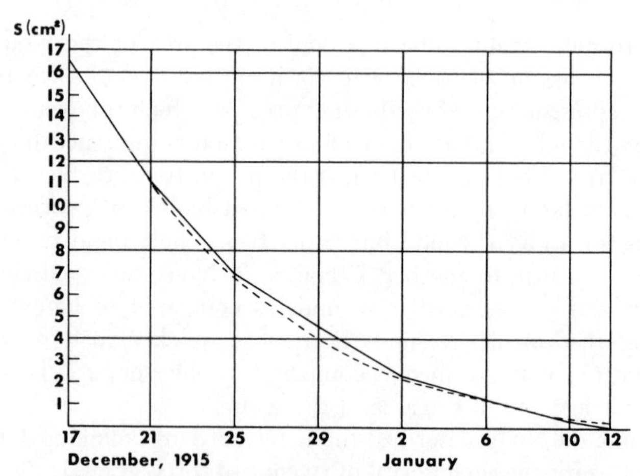

FIGURE 2. Relationship between areas of the wound and time. Solid line: observed values; dash line: calculated values. In Lecomte du Noüy, *Le temps et la vie* (Paris: Gallimard, 1936), p. 109.

is characterized by a major slowing down of the healing process as the same k applies for a 150 cm^2 wound in a 20-year-old man and for a 50 cm^2 wound in a 40-year-old man. Cellular wound repair enables us therefore to assess a coefficient k, proportional to the physiological age, which can be used as a *measure of aging*.

It should be pointed out that relation (*1*) is not homogeneous with respect to the dimensions. Namely the same coefficient k should have the dimension T^{-1} and $T^{-1/2}$ to fit into the equation. This difficulty can be overcome by the use of two coefficients k and k' with the respective dimension of T^{-1} and $T^{-1/2}$ and with very close numerical values, inasmuch as later reinvestigation of the wound-healing law led Lecomte du Noüy to introduce more than one proportionality factor.

Namely, the simple exponential function:

$$S = S_o e^{-kT} \qquad (2)$$

where S_o = initial wound surface

T = age of the wound

k = proportionality factor (T^{-1})

is not satisfactory, as k increases regularly for different times of T. He therefore changed this relationship to the general formula:

$$S = S_o e^{-k(T + T^2/2p)} \qquad (3)$$

where k = proportionality factor (T^{-1})

p = proportionality factor (T)

The time of complete healing can be then calculated by solving the equation for $S = 0.4$ as a wound of 0.4 cm^2 heals in twenty-four hours.

As to the biological meaning of equation (*3*) Lecomte du Noüy relates the exponents to two different biological processes, namely to granulous contraction of the wound for:

$$S = S_o e^{-kT}$$

and to epidermisation for the factor $T^2/2p$

These facts led Lecomte du Noüy to consider the concept of time, of physical time, without beginning or end, which passes at a uniform rate without acceleration or slowing down. But there is also a physiological time, which starts and ends with us and which affects differently physiological processes, depending upon whether we are in childhood or in senescence.

Woundhealing is work, and the rate of the work is high at the beginning of life and slows down later on. These rates, measured in physical time units, imply that different time intervals are needed to do the same work at different ages. Lecomte du Noüy writes that we have in ourselves a machine to register the time; our subconscious gives to our intelligence the raw information: time passes slower and slower with aging.

Our internal physiological time does not appear to us to pass at a constant rate as measured with external physical time. If a period four times as long is required to heal the same wound at fifty years as compared to ninety years, this means that physical time will flow four times faster as compared for a child of ten years. "There is a physiological time which has only significance for organisms which are born, grow old, and die normally. The only time which is of importance for man is his own time elapsing between a prawn and a tomb."

Aging is often considered as being the result of the wearing out of tissue, of the accumulation of toxic products, of progressive alteration of functions. The understanding of the aging process is essential if one wants to try to influence it. We have tried to change biological time by the following procedure. Figure 3 shows the weight gain of the male Wistar rat on a free and equilibrated diet. The growth curve is sigmoid and can be expressed by the relationship:

$$P = [485^{1/3} - (485^{1/3} - 7^{1/3})e^{-0.0156\,t}]3 \qquad (4)$$

where t = time in days.

This equation enables us to assess accurately the age of the animal during the whole active growth period. But if the dietary intake is restricted in such a way that growth is practically arrested during 95 days, the weight can no longer indicate the age of the animal as at the end of the restriction; the 140-day-old animal has the weight of a 48-day-old rat, but if now the rat is allowed to eat as much as he wants, what will happen? One of two things may happen. If the passage of time has been effective, growth will resume—but with the intensity of the animals of the same age; weight gain will be slow as compared to the gain at the moment at which food was restricted. On the other hand, growth could resume with the same intensity as at the time of interruption. The flow of time would then have left

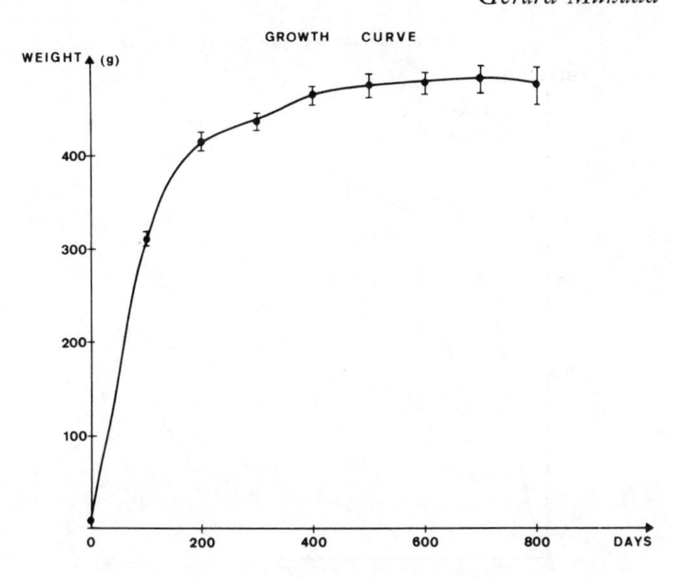

FIGURE 3. Body weight curve for eight rats; standard errors are shown.

the growth potential unaffected and the rat would not have aged. Due to the similarity of the processes of growth and cicatrization, this would indicate that physiological time has not passed during the period of restricted food intake for as long as the equivalent of five years in term of human life.

Let us see what will happen. The growth rate is the same as at the time growth was arrested (Figure 4). *Physical time* has passed; *physiological time* has stopped. This is proved to be true by taking the weight as a parameter. Will it still be true if one investigates the behavior of a system very sensitive to age—the skeleton? During the growth period, skeletal age can be assessed so precisely that the bone can probably be considered as the biological clock of the body. Bone is continuously renewed. During growth, anabolic processes are more important than catabolic processes. During adult age, the two processes match each other. During aging, catabolism predominates over anabolism, and bone loses calcium and becomes fragile. Senile osteoporosis is probably the most common disorder of age since it expresses the aging process of bone.

In his book *Time and Life*, Lecomte du Noüy contrasts the behavior of cells in tissue culture, which neither age nor die, to the behavior of the whole organism, which has different repair rates during its evolution. The passage of time affects differently tissue culture and the intact animal, which is driven to death. It has to be observed

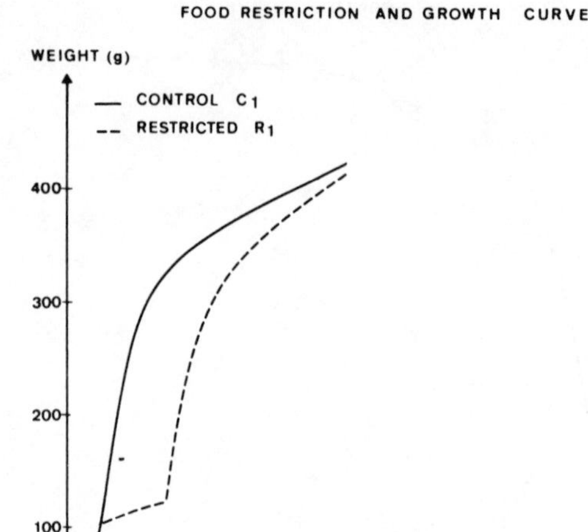

FIGURE 4. Body weight curve of control and food restricted rats.

that the experiences of Carrel concerning the immortality of diploid cells have been critically reviewed. It is no longer accepted that a diploid cell can undergo an unlimited number of cell divisions. If a graft from an old animal is put into a young animal, the graft will survive the donor but will die before the recipient, as if the life span of the graft would be limited. But even if Carrel's conclusion should prove to have been wrong, this would only imply a generalization of the concept of biological time discovered by Lecomte du Noüy.

Cell differentiation, organogenesis, and growth lead to aging and death, which are part of the program which was given to the organism at birth. Aging is not a reversible process, as it is equivalent to growth and cell differentiation. A unique program seems to direct all the steps of life, and aging starts with the first cell division and not with the first gray hair or wrinkle. Aging is enhanced in a young organism, in which the program is read rapidly. Claude Bernard defined life by the simple sentence, *La vie, c'est la mort.* Diseases and death are included in the life program, the coded tape carries at a given place the information which will be read as heart attack or cancer leading to death. Death is a mechanism essential to the preservation of the species. Death of the actual generation gives a better chance to the coming ones by suppressing the struggle for life between these generations. Bergson wrote that life appears like a stream flowing from germ to germ by means of a developed organism. As human beings we are mostly interested in the developed organism, whereas for the species, it is the stream that counts.

Now the question arises: What is the coded tape at the cellular level? Nucleic acid could be ordered along a tape, which would be read from beginning to the end of life. In order to move from one nucleotide to the next, the cells should have first completed the instruction of the first nucleotide. Regulation loops can be worked out to make the tape move a punch ahead, and a new nucleotide sequence becomes operative. Such a mechanism operates with time and is equivalent to a clock. This type of clock has to be on time, as for instance puberty starts at an age which does not vary too much from one individual to the other, and as the events follow always in the same order with a negligible percentage of error. And even the errors—like Turner or Klinefelter syndromes—result from chromosomal aberrations. The reading is made according to an altered program rather than the misreading of a normal program.

Are the clocks present in all the cells of the individual or do they confine themselves to given organs only? In mammals we could expect that the nervous system would play an important role as a remarkable integrating system. It could store the steps of the cellular life, of growth, of adulthood, of aging and play the role attributed by Lecomte du Noüy to our subconscious.

Under such conditions, is it possible to rejuvenate, to reverse biological time? This would mean the backreading of the coded tape and imply the possibility of reversing one after one all the steps of development to the stage of the first cellular divisions. This hypothesis seems wrong, and the logical conclusion would be that rejuvenation is impossible.

If it is impossible to read the coded tape backwards, it should be possible at least to slow down the rate of the reading and maybe to inhibit the acceleration of biological time. The length of youth would be prolonged, the evolution to senescence would be slowed down, and the program leading to death would be read slowly. These considerations have a profound bearing on the studies related to aging. It is compulsory to do something to modify this process as long as the individual keeps enough potentialities. The aim is to slow down the speed of evolution. Something should be done before the signs of aging appear, if one wants to modify the time course of the aging process. When the evolution potential is exhausted, when the coded tape comes to the last punches, we cannot any more slow down a process coming to its end.

Geriatrics can describe accurately the events of the third period of life, but it cannot possibly have any practical consequences as to the aging process itself. Geriatric medicare takes place too late. The play is close to the end, the curve is completely described, and the evolutionary potential is therefore reduced.

Schrödinger describes the living organism as being able to drink negative entropy. I would like to add that the way of drinking is of the utmost importance: the rate will vary spontaneously during a lifetime, and the life elixir may consist in regulating the rate of drinking. This would go along with the law, discovered by biophysicists, which links the length of life time (L) with the intensity of metabolism (I) for several animal species:

$$L \times I = constant$$

We have to decide for ourselves how we will spend the life capital, the potential of evolution, if we prefer, as Chamfort said: *Etre passionné et vivre ou être raisonnable et durer.*

On Time, Information and Life

Costa de Beauregard

Time, from a physicist's point of view, has two major aspects. First, it is a "measurable magnitude," as physicists say in their jargon. That is, the addition of two successive time intervals can be validly defined. To relate how this has been done would be a long story, including a recall of what universal laws, universal constants, dimensional equivalence of magnitudes and equivalence coefficients mean to physicists.

Suffice it to say that the first epoch-making definition of time as a measurable magnitude is through the well-known Galileo-Newton formula $F = m^{\infty}$ for moving point particles. This formula relates in a universal way the physical magnitudes of time, space, force, and mass. The latter three being directly measurable, time is thus defined as indirectly measurable. The technological breakthrough corresponding to this scientific achievement has been Huyghens' pendulum and escapement mechanism, still in general use today, which has produced the revolution of precise timing in everyday life and also the possibility of knowing longitudes at sea.

The second epoch-making definition of time as a measurable magnitude is through Einstein's theory of relativity[1] and Minkowski's formula for space-time intervals. The time interval is thus directly related to the space interval with the velocity of light *in vacuo* as an equivalence coefficient. This is casting into precise scientific form Aristotle's very appropriate statement that "we measure time by means of motion, and motion by means of time."[2] And here also we have a technological breakthrough. As the physics underlying the space-time four-dimensional geometry is the wave propagation of light, to use a light wavelength as a length standard and a light period as a time standard is, so to speak, relativity in act, and makes c a universal constant by definition.[3] In fact, it happens that this is today's joint status of high precision metrology and chronometry.[4]

So much for time as a measurable physical magnitude.

The second major aspect of time in physics is the irreversibility occurring whenever the energy balance implies heat—that is, in such commonplace phenomena as friction or such unsophisticated devices as brakes as well as in the much more sophisticated heat engines obeying the famous Carnot-Clausius law.

Starting through deep insight from these humble (or at least technological) origins, the development of the theory of irreversibility has turned out to have extremely profound implications—implications that even today open avenues for exploration. *Entropy*, this "incredibly abstract concept" coined by Clausius has been interpreted, at the turn of the century, by Boltzmann and Gibbs as the logarithm of a probability. Quite recently, the new information concept (which was truly implicit in its early forms) has been recognized as basic in all technological and scientific applications of probability theory. Scientists have had to learn that whenever the kin concepts of probability, entropy, or information are at stake, physics comes very near indeed to metaphysics. It thus turns out that the irreversibility problem has something to do not only with the working of cybernetic machines, but also with ontogenesis and phylogenetics in biology, and with behavioral problems in animal organisms. Such general metaphysical schemata as causality and finality are touched upon in the course of the inquiry—and that is why I happen to be speaking on physical irreversibility at this Notre Dame meeting about Lecomte du Noüy.

The important point, which has been fully recognized only recently, is that physical irreversibility is never deduced, but truly *postulated* at the very root of any theory dealing with it.

Consider, for instance, phenomenological thermodynamics. The two facets in Carnot's postulates are:

1. That heat will flow from a place at high temperature to a place at low temperature, not the other way.
2. That in a monothermal situation work is convertible into heat, not the other way.

If the time arrow were reversed in these two Carnot postulates, it would also be reversed in all the rest of phenomenological thermodynamics. Thermodynamical irreversibility is thus not spontaneously generated in the course of deduction, but is built into the so-called second law at its very root. It remains true, of course, that phenomenological thermodynamics has uncovered a wonderfully large class of irreversible phenomena in every branch of physics.

Speaking of trying to reverse the time arrow in Carnot's postulates immediately raises very difficult questions. Assuming a reversal, it would be dangerous, says Grünbaum, commenting on Poincaré, to get into a lukewarm bathtub, because one could never tell which end was going to boil and which to freeze. Also, if friction were an accelerating rather than a damping process, setting bodies in any direction and magnitude of motion would make bowling a dangerous game indeed. Poincaré, commenting upon such examples, concluded that the anti-Carnot world would be a lawless world. The exact point, however, is that our usual Carnot world is such as to allow physical prediction rather than retrodiction, while the anti-Carnot world would allow retrodiction but not prediction. In this sense, our usual Carnot world may well be called a causal world, while the paradoxical, anti-Carnot world should be called a final or teleological world.

We have previously recalled that thermodynamic irreversibility has been reinterpreted by statistical mechanics as the tendency to increase probability which is displayed in all physical arrangements or devices. Everybody believes that shuffling will destroy any kind of order imparted to a deck of cards, but nobody will rely on shuffling for ordering the deck.[5]

That probability tends to increase and not to decrease with time (at least in physics) is a trivial fact, but this does not render obvious the reasons for this fact. For if the probability for exchanging two cards in a deck only depends on their positions, the basic process is time symmetrical. Then if one understands quite easily that shuffling will destroy order in the future (that is, understands statistical prediction) one cannot explain so easily why the "blind retrodiction" that the ordered deck has emerged from shuffling is in fact unbelievable.

As early as 1763 Bayes produced a rule for dealing with such problems, and, as we shall see, this rule has far-reaching implications. Underlying Bayes' rule is his formula for conditional probability

$$p_i' = \frac{q_i p_i}{\Sigma q_j p_j}, \tag{1}$$

where the p_i's denote the intrinsic probabilities of physical occurrences, that is, those depending on the internal dynamics of the system under study, and the q_i's extrinsic or conditional probabilities. The rule then states, in Watanabe's words, that *blind statistical prediction is physical while blind statistical retrodiction is not*; that is, for prediction, equal q_i's, entailing $p_i' = p_i$, will do, while any sound retrodictive estimation will imply an *a priori* estimation of the extrinsic q_i's. Of these only one general thing is said: they must not be equal, because their equality would mean blind retrodiction.

That this is the general state of affairs in physics has been expressed by Willard Gibbs in a solemn sentence:

> It should not be forgotten, when our ensembles are chosen to illustrate the probabilities of events in the real world, that while the probabilities of subsequent events may often be determined from those of prior events, it is rarely the case that probabilities of prior events can be determined from those of subsequent events, for we are rarely justified in excluding the consideration of the antecedent probability of the prior events.[6]

So speaks, ever since the old days of Bayes, the oracle of probability theory. But this amounts to saying that irreversibility in probability theory simply does not spring from nowhere, because it is truly built in the Decree from Above: blind retrodiction forbidden.

This conclusion is precisely isomorphic to the one we reached when speaking of phenomenological thermodynamics. Thus it is true that in the general theory of probabilities blind predictability is synonymous with the existence of causality, while the prohibition of blind retrodiction in physics is synonymous with the exclusion of teleology. Whence a traditional denomination for Bayes' principle of probability of causes.

As for the more specific kinds of irreversibility dealt with by statistical mechanics, suffice it to say that the Clausius concept of entropy ("a wonderfully abstract concept," as Poincaré wrote) has been interpreted by Boltzmann and Gibbs as the logarithm of a probability, so that, provided that the basis of logarithms is greater than one, the principle of entropy is a mere specification of the general principle of probability increase.

In other words, as recognized by van der Waalsin in 1911, the statistical interpretation of the Carnot-Clausius principle is merely a specification of the Bayes principle. This is the concise answer to the well known Loschmidt and Zermelo paradoxes.

Trying to find out the physical motivations of the Bayes or the Carnot decrees will provide further insight.

To say that card shuffling alone will not produce a low probability deck of cards is to say that some external occurrence has to intervene. To say, for instance, that an ink drop will diffuse but never concentrate in a glass of water is to say that it came out of a pipette. So the physical interpretation of Bayes' extrinsic coefficients clearly is the representation of the interaction out of which the system under study has been segregated. And to say that these extrinsic coefficients have to be used in retrodiction rather than in prediction amounts to saying that physical interactions develop aftereffects and not before-effects. For instance, introducing the pipette in the glass containing ink diluted in water and sucking will never concentrate and extract the ink.

Statistically speaking, there is a very definite connection between the principle of retarded actions and the principle of increasing probability. Any one of the two can be deduced from the other. In so doing, one enlarges one's scope to the total system comprising two interacting subsystems.

Following this scheme (which Reichenbach and Grünbaum call the "theory of branch systems") one will eventually be led to the consideration of the whole universe, and to the conception that if all partial evolutions display the same Bayes or Carnot time arrow it is because they are not truly isolated from each other, and thus they reflect[7] the physical irreversibility of the whole world. In this sense, the Bayes principle is interpreted as a coercive influence of the whole on its parts. But this, of course, is not explaining why the whole universe is going this way and not the other, or rather (as today we must think in relativistic terms) why all of us living beings are going through space-time in the direction where probability is increasing, not decreasing.

To this question we will come back later. Suffice it to say for the moment that establishing a correspondence between increasing (resp. decreasing) probabilities and retarded (resp. advanced) interactions gives more consistence to their one-to-one connection with causality (resp. finality).

So it definitely turns out that in the statistical theories with which we are dealing, the irreversibility principle is not at all formulated as an intrinsic elementary law of evolution,[8] but rather as an extrinsic global fact. As Mehlberg puts it, irreversibility is of a factlike rather than lawlike character. Speaking more technically, Bayes' principle is a boundary condition imposed upon the integration of the statistical equations: a boundary condition stating that the macroscopic kind of evolution is retarded (not advanced) action—that is, causality (not finality).

But there is in physics another very general irreversibility principle, also stated in the form of an arbitrary decree excluding just half of the a priori conceivable evolutions, and also expressed in the form of a boundary condition: the principle that physical waves are retarded, not advanced. This principle states that active sources of waves do exist but active sinks do not. May it be that a physical connection exists between the two principles of increasing probabilities and of retarded waves?[9]

Asking such a question immediately calls to mind the existence of many examples pointing towards a positive answer. Consider for instance the slowing down of a meteorite in the earth's atmosphere. The fact is that in this case the mechanism of entropy increase is mediated by the emission of retarded ballistic and light waves. More specifically, consider the case where, between times t_1 and t_2, a physicist moves a piston in the wall of a vessel containing a gas in equilibrium, and finally brings it back to its first position. The fact is that Maxwell's velocities distribution law is altered after time t_2, not before time t_1; also, that the perturbation is propagated as a retarded wave emitted, not as an advanced wave absorbed by the piston. Speaking on purely observational grounds, there is a strong hint in favor of the conjectured connection.

But it obviously remains impossible to formulate mathematically such a connection as long as one is unwilling to use a theory of waves implying some kind of statistics. Today, with the advent of such concepts as phonons or rotons it seems that all conceivable physical waves, including those propagating in material media, should be quantized. And this by itself brings in the statistical element we need.

My present purpose, however, is not to actually carry out such an ambitious program. I merely wish to show, using very specific examples, what I precisely have in mind.

Before coming to this I will recall two very significant hints that the development of quantum theory has produced.

In 1908–1909 a lively controversy had opposed Ritz to Einstein. Ritz believed that the principle of retarded waves was a tacit assumption when deducing the second law of thermodynamics, while Einstein maintained that it was the principle of probability increase that was underlying the law of wave retardation. Had Ritz and Einstein known that the quantization of light waves was only half of the truth they needed and that de Broglie's wave mechanics was to come, then they would have known that scattering of particles (in the sense of statistical mechanics) is also a scattering of waves. Certainly both of them would have realized that they were saying the same thing in a reciprocal form.

Also, Planck's definition of the entropy of a light beam, where the h constant occurs and the photon concept is thus implied, has the consequence that the entropy increase in coherent light scattering or diffusion. But coherent wave diffusion is the concept associated with retarded waves, while coherent wave confusion would be the (paradoxical) one associated with advanced waves. So we have here, in the realm of the Bose statistics, an instance where there is a one to one correspondence between probability increase and wave retardation.

These two historical occurrences make a good introduction to the point I now make.

In the theory of quantized waves, and already in the wave mechanical form of quantum theory of the 25's, retarded waves are used in statistical prediction as advanced waves should be used in blind statistical retrodiction. It thus seems quite clear that, as soon as we are dealing with quantized waves, the physical nonexistence of advanced waves (on the macroscopic level) and the prohibition of blind retrodiction are merely two different names for one and the same principle. This is the point I wish to discuss by using specific examples.

Consider first the case of a plane monochromatic wave falling upon a plane grating, the wave planes being parallel to the lines on the grating. A finite number g of plane waves is reemitted from the grating, and we assume for simplicity that they all have equal intensities.

Now, any of these g outgoing waves may be excited by any one of the g incident plane waves belonging to one and the same family (which of course comprises the incident wave we spoke of first).

In terms of quantized waves, the grating thus induces transitions between two sets of g orthogonal states, and we have assumed for simplicity's sake that all the transition probabilities are equal.

Now we assume, for simplicity's sake again, that the quanta, or particles, are traversing the grating with mean time intervals long enough for being individually recognizable; this is to say that they do not belong to the same wave trains or quantal cells, so that we are dealing with classical, not quantal statistical techniques.

It is well known and easily found out, that the number of ways in which a given distribution of n_i particles in the g_i cells is obtainable is

$$P(n_i) = \frac{n!}{\pi\,(n_i!)} \qquad (2)$$

a number which increases if one picks a particle in a cell and puts it in a more occupied one. Thus, the most probable distribution is (as was intuitively felt) that one in which all n_i's are equal (with eventual difference of $\pm$ 1).

Finally we consider two symmetrical experimental arrangements.

First arrangement: All incident corpuscles are carried on one incident plan wave; they come out of the same collimating source. Then, blind statistical prediction yields the result that the corpuscles emerging from the grating are equally distributed among the g outgoing plane waves. But this is the result that the principle of retarded

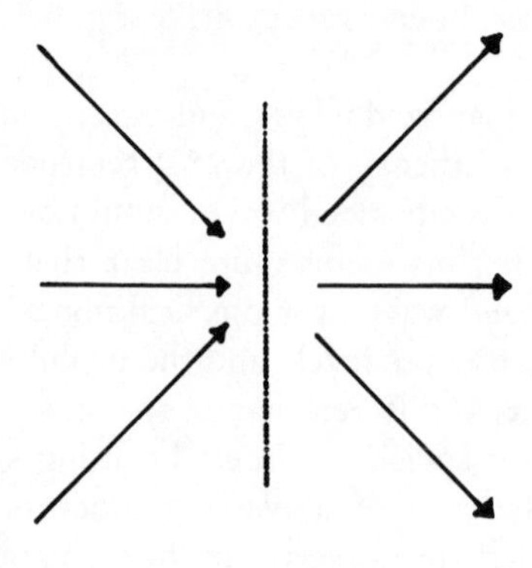

FIGURE 1.

waves yields also: phase-coherent scattering or diffusion occurs, with equal intensities of the outgoing waves (under our simplifying hypothesis).

Second arrangement: All corpuscles which are detected as emerging from the grating are carried on the same outgoing plane wave; they are received in the same collimating sink. Then, blind statistical retrodiction would yield the very suspect conclusion that these particles, when falling upon the grating, were equally distributed among the *g* possible incoming plane waves. But the same conclusion would follow from the unphysical principle of advanced waves, then stating that phase-coherent confusion instead of diffusion is occurring.

In other words, it is clear in this case that blind statistical prediction and wave retardation are synonymous—just as, on the other hand, the prohibition of blind statistical retrodiction and the macroscopic nonexistence of advanced waves.

The preceding example belongs to a very wide class of similar ones, which can be dealt with all at once by using von Neumann's theory of irreversibility in the quantum mechanical measuring process. Though the significant formulas can be cast in a very simple form (as I have explained elsewhere) I will not go into them. Suffice it to say that a mere rewording of von Neumann's proof shows that the probability increase directly follows in it from the tacit assumption that the measurement procedure is a source of retarded waves rather than a sink of advanced waves. Had the opposite, paradoxical assumption been made, then the opposite, paradoxical conclusion of probability decrease would have followed.

It is then obvious that the tacit assumption underlying von Neumann's irreversibility proof is the Bayes principle which is there equivalent to the principle that the macroscopic wavelike emergence from quantal processes is of purely retarded character.

Things being thus, it is appropriate to discuss the special traits assumed in the present case by the time symmetry inherent in the elementary evolution laws. It would seem at first sight that there is nothing more in them than the obvious symmetry properties of the so-called spacetime Green's functions. But it happens that Einstein's sagacity has uncovered at the 1927 Solvay Council a family of very vexing paradoxes, the best known one being the 1935 Einstein-Podolsky-Rosen paradox. The version we are devising here will bring out the crucial points.

Suppose a monochromatic beam is split by a semitransparent mirror into two beams, (*a*) and (*b*) [Fig. 2]. Speaking of particles, there is for each a probability that it is transmitted and a probability that it is reflected. Assume for simplicity's sake that there is just one particle on the incident beam. If an observer O_a operating on the beam (*a*) either finds that the particle is present or absent,[10] then he knows that it is respectively absent or present on the beam (*b*); that is, he knows that the result of a similar measurement performed on the beam (*b*) by an observer O_b. Now, the distance between observers O_a and O_b may be extremely great; moreover, speaking in terms of relativity, it can be spacelike. So, the logical inference that either O_a or O_b is able to make is neither prediction nor retrodiction, but truly "telediction."

While everybody feels that there would be no miracle in this if one simply could hold a deterministic view of the world, things are

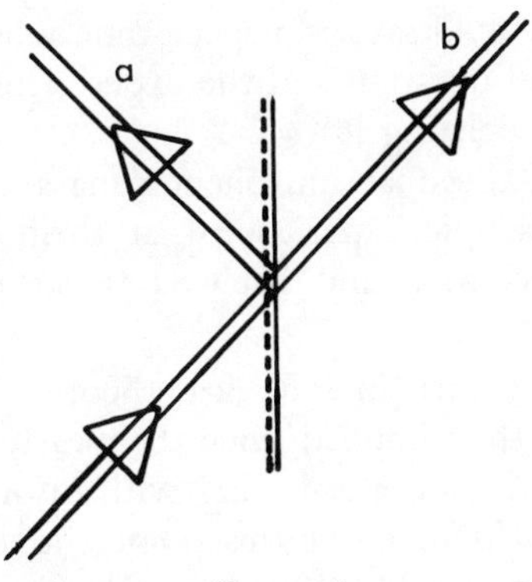

FIGURE 2.

not so simple with a fundamentally probabilistic one—and even less with the one afforded by quantum theory.

The first question is: along which channel is the logical inference telegraphed? The answer of the calculation is quite explicit: the "telediction" goes in space-time along a Feynman style zigzag with its apex in the space-time domain where the die is cast, first towards the past and then towards the future. Thus, the first conclusion to be drawn from the Einstein paradox is that the previous Einstein prohibition to telegraph into the past is removed at the level of an elementary quantal process. It was a macroscopic, not a microscopic prohibition.

Shall we then get rid of the paradox by merely stating that, according to the relativity theory,[11] Minkowski's space-time has just the same kind of physical reality as Euclid's space had? Old probability problems with balls in boxes have never suffered from paradoxes owing to their spatial description. If we could just say that, being amplified by some triggering device, the presence or absence of the particle on beam (*a*) or (*b*) is a macroscopic, objective, event, which as such is written once and forever in Minkowski's space-time, would not everything be all right?

There is one point, in the quantum theory, that prevents the preceding discourse from telling the whole truth, and it is that performing a measurement contributes to producing the result of it. In the present case, if no determination of the occupation numbers of beams (*a*) or (*b*) were performed, then they would remain phase-coherent and their probability amplitudes would still be able to interfere; but this is no longer the case if either occupation number is measured. In other words, in the quantum theory acquiring knowledge entails actuating the system under study, and this makes the sting of the Einstein paradox. Bohr's statement that it is not the last critical stage in the measuring process, but the experimental arrangement as a whole (I would say, as a space-time whole) which makes both the experimental question and the experimental answer softens the paradox without deleting it.

My feeling is that the quantum phenomena are so much outside our ordinary macroscopic experience that their deep implications are not yet truly understood, and that we must keep our minds open.

The latest dramatic turn in statistical theories has been the full recognition of what they implied since their early days in the time of Pascal and Fermat: they are dealing with information. So many important names are involved at this point and in so many disciplines that listing them would seem a litany. Suffice it to say that for technical reasons, information has to be an additive magnitude, and is thus defined as minus the logarithm of the probability. That is, its formal definition is the same as that of a negative entropy or (as Brillouin says) a negentropy.

Shall we then bluntly say, as Boltzmann wrote somewhere, that "entropy merely is missing information," or with Lewis that "gain in entropy always means loss of information, and nothing else"? Is it a subjective concept? Is it not true that the sun would radiate also if there were on earth neither scientists, engineers, nor men in the street, with their more or less limited understanding of the phenomenon? The point is, however, that it is this very understanding that enables them to build dams or pump oil; that is, to recapture part of the dissipated energy flow. So if there is truth in these statements by Boltzmann and his followers, there is also something very misleading in them as long as one is not aware that information is a twofold concept: knowledge, and power for intervening. In this sense information is definitely not a purely subjective concept; it is an indissolubly subjective and objective concept.

At this point one should remember that the everlasting struggle between objectivists and subjectivists in probability theory has never come to an end, so the wisest position may be to hold that the probability concept is neither objective nor subjective, because it is by its very nature indissolubly objective and subjective. Consider the disorder concept, of which Schafroth writes: "Scientists exist who pile up papers in a seemingly random fashion on their desks, yet know all the time how to find a given thing. If someone brings apparent order to this desk, the poor owner may be unable to find anything." The point that must not be overlooked is that a fine-grained knowledge of what is contained in the piles is the sine qua non condition for being able to put the papers in true (not apparent) order.

It has turned out in physics also that the entropy concept is somewhat sphinxlike. A nineteenth century locomotive engineer, dealing with macroscopic problems, certainly did not doubt that entropy was an objective concept. But the analyses by Smoluchowski, Szilard, and their followers have clearly shown that whenever one asks such questions as "Where is the one molecule assumed to be in a box after a partition is inserted?" Entropy is indeed lack of knowledge, and nothing else. Thus the entropy concept has two limits: an objective one in macroscopic physics and a subjective one in microscopic physics. There should then be no wonder that the same is true with physical waves: classical macroscopic waves certainly are very objective in their behavior, while on the quantum level these very same waves have a strongly subjectivistic flavor—their so-called wave collapse, for instance, duplicating the information jump inherent in any statistical test.

The important point that the possession of some fine-grained knowledge allows its owner to produce coarse-grained order is known, in physics, as the Maxwell demon problem, and has been thorougly discussed by Smoluchowski, Szilard, Demers, Brillouin. It turns out that, according to certain rules we will come back to, fine-grained knowledge is convertible into coarse-grained order. Thus, if Maxwell's demon has a small light bulb enabling him to recognize the individual molecules, he will then be able to select from air inside a vessel all nitrogen molecules on one and all oxygen molecules on the other side of a partition.

The rules I was alluding to are what Brillouin calls the generalized Carnot principle. Brillouin and others have very convincingly shown how the information I gained in some physical measurement is in fact borrowed from the preexisting negentropy N_1, which is conveniently symbolized as

$$N_1 \rightarrow I, \tag{3}$$

and that the irreversibility law

$$\Delta N_1 \geq \Delta I \tag{4}$$

then holds. As Gabor puts it, "one cannot have anything from nothing, not even an observation." Thus acquiring experimental knowledge is not costless, as was implicitly believed in old days. The reason why such a fundamental law has escaped previous analyses is as Brillouin points out, the very smallness of the universal constant $k\,Ln\,2$ converting an entropy expressed in practical thermodynamic units into a bit of information expressed in the natural binary units according to the formula

$$\Delta N = k\,Ln\,2\,\Delta I. \tag{5}$$

This simple fact contributes to the whole flavor of everyday life and explains how we scientists are having preprints and reprints sent to our colleagues, who eventually throw them right into their wastebaskets. It is only when very high precision is required that the negentropy cost of the experiment may become prohibitive.

On the other hand, converting fine-grained knowledge into coarse-grained order, as Maxwell's demon is supposed to do, may be summarized in the formula

$$I \rightarrow N_2 \tag{6}$$

with then, as Demers, Brillouin, and others have shown, the irreversibility law

$$\Delta I \geq \Delta N_2 \tag{7}$$

This time the money-change rate plays in the wrong way: producing negentropy costs a loss in information units. This I am well aware of when struggling to keep my office in order.

On the whole, the generalized Carnot principle of physics[12] obeys the formula

$$\Delta N_1 \geq \Delta I \geq \Delta N_2, \tag{8}$$

the older Carnot principle being obtained by simply erasing the middle term. Here Brillouin writes triumphantly that Maxwell's demon has been exorcized because Carnot's principle is safe. Even more, Maxwell's demon might be dismissed just as well, because if we are merely using the coarse-grained negentropy corresponding to the energy stored in the battery of the light bulb to produce the smaller coarse-grained negentropy corresponding to the oxygen and nitrogen being separated, then a cyberneticized robot will do just as well. It is true, however, that such a robot has first to be conceived and then built, and this is another story.

Before we proceed in the analysis of our problems, I must indicate that this is the kind of procedure certainly going on in phylogenesis, ontogenesis, and even behavior of plants, animals and men. Consider, for instance, the problem of reproducing a given item of information: drawing copies of a text or a photo, or drawing copies of the genetic code in cell multiplication. Also, consider the case in which a professor gives a lesson to a whole class. It is true that part of the professor's information is lost by being absorbed in the walls of the classroom and by going into the ears of distracted students. Nevertheless, at the end of the lesson the professor has not lost his information while many students have gained most of it. How is this possible? Well, the fact is that, the larger the classroom, the louder the professor has to speak, and it then becomes clear that all the extra information is borrowed from a preexisting negentropy; the professor must have had some rest, and maybe breakfast before entering the classroom. Similarly the photocopy machine works with a light bulb emitting retarded waves. And (in some complex manner

I am not aware of) the copy of the genetic code can only go on if negentropy is fed into the growing organism. Discussing this would bring us back from animals to plants, and to photosynthesis, and finally to the sun lavishly emitting photons.

Everything is very true in this discourse but by relying only on de facto rather than de jure considerations we are missing the point. If it is a priori assumed that Maxwell's demon is so radically dependent on ordinary means of observation and action, then there is certainly no point in having him exorcised. What we must not forget is the observation made by Mohlberg and others, that physical irreversibility is factlike rather than lawlike. If this is a general statement—and I do believe it is because I independently made it—then it is true here also; that is, in the generalized Carnot principle. In other words, it must be a de facto rather than de jure law that makes the learning transition $N_1 \rightarrow I$ so much easier than the acting transition $I \rightarrow N_2$.

It seems to me that it is this de facto situation which is reflected in the very smallness of Boltzmann's constant:

$$k \simeq 1.38 \times 10^{-16} \text{ c.g.s. units.}$$

It must not be forgotten that the very choice of our practical units reflects our situation-when-being-in-the-world (as existentialism puts it). Taking another example, if the universal constant c of relativity theory, the light velocity *in vacuo*, is so large when expressed in practical units—say, centimeters and seconds—this is because we are so built that centimeter and second are for us practical associated units of space and time. The velocity of our nervous influx is of a few decimeters per second, and I truly wonder what our social life would look like if it were instead of the order of the light velocity. Relativistic kinematics would then be the expression of common sense, and you and I would not feel that we are living in this room at the same time. The de facto smallness of Boltzmann's constant certainly has quite analogous implications—implications that it is the task of cybernetics to unfold. Incidentally, Planck's constant h is also a very small universal constant; and so it turns out that all twentieth century physics is concerned with very small or very large universal constants. This is merely stating that it is fundamentally concerned with phenomena lying much outside the realm of everyday experience.

A time honored exercise in both quantum and relativity theories is to let h or $1/c$ go to the limit 0, and see how various prequantal or prerelativistic conceptions are thus recovered. So let us try to see which is the precybernetic conception which is recovered in the limit $k \to 0$. Returning to the equivalence formula

$$\Delta N = k \, Ln \, 2 \, \Delta I,$$

we see that this would make observation costless and action impossible. But this is a very well known conception; it has been defended as the "theory of phenomenal consciousness."

In cybernetics, consciousness is no more an invited spectator; she has to pay for her ticket. But this, as we will now make clear, is the very condition allowing her to become an actress also.

What is intended now is to examine the implications of complete time symmetry of the transition formula (5), that is, of the complete de jure symmetry between the learning transition $N \to I$ and the action transition $I \to N$.

It should be noted at this point that Aristotle–the promoter of the information concept and appellation–considered that information could be both an acquisition of knowledge and an organizing power. Since then, a de facto situation has tended to prevail, where the first acceptation has wide recognizance, while the second one was restricted to a small circle of philosophers and thinkers–mainly those very few interested in finality. It is thus a notable fact that cybernetics, without having searched for it, has hit upon the de jure symmetry of the two processes of acquiring a knowledge and producing order–that is, it has rediscovered the twin faces of the Aristotelian information concept.

Now, as we have seen, the somewhat passive process of learning from observation belongs to a class the limit of which is the completely passive Carnot process of entropy increase in a closed system with no work produced. Generally speaking, the learning transition $N \to I$ can occur in the entropy increasing evolutions, that is also (as we have seen in some detail) in the retarded actions or retarded waves phenomena; on the whole, within the wide class of phenomena obeying blind predictability, that is, macroscopic causality.

Symmetrically, the process of producing by will some kind of order occurs in the form of the active transition $I \to N$, that is, in entropy decreasing evolutions, advanced actions or advanced waves phenomena and, on the whole, within that restricted class of phenomena obeying blind retrodictability, that is, teleology or finality.

Summarizing, willing awareness and learning awareness should respectively occur within progressing and regressing statistical fluctuations. Also, cognizance and causality on the one hand, will and finality on the other, are two symmetrical pairs of associated concepts, so that there is no wonder that finality is so elusive when observation and cognizance are the research methods employed. For it is by their very nature, or by definition, that causality is the kind of process evident to cognitive awareness, just as finality is the one obvious to willing awareness. The queer point is that it is only very recently that the connection between entropy increase and causality has been fully recognized by quite a few authors (Reichenbach and others), while the symmetrical connection between entropy decreases and finality was understood long ago.

That cognizance and will awareness are equally genuine[13] may be taken as the subjectivistic aspect of the de jure reciprocity in (coarse-grained) negentropy and information conversions, of which the over-all symmetry between the sensitive and motive nervous chains is an illustration (quite apart from the de facto information loss along the lines). So the de jure question we are raising beyond the de facto situation that observation is easier than action (which is the expression of the irreversibility principle in the present context) is that there might be after all information sources of nonexperimental origin. Denying this would be denying that anything really new comes out in our universe, as then any order should proceed from some antecedent larger order. While this latter certainly is the de facto situation, I feel it would be distressing to believe that it is the de jure and exceptionless state of affairs.

In conclusion, I would like to say that finality should be taken no less seriously than the position. For the position (which is de jure symmetrical to the trivial electron or negaton) is de facto much rarer than it. The position concept was an open possibility in the mathematical formulas of Dirac's theory of the electron, and its experimental discovery by Anderson has thus filled an open theoretical compartment. It happens also that both theoretical and experimental physics have learned quite a lot from the position. Why not the same with the finality concept?

By way of postscript I wish to give two quotations from physicists. One is from Léon Brillouin, at the end of his fine book, *Science and Information Theory:*

> Relativity theory seemed, at the beginning, to yield only very small corrections to classical mechanics. New applications to nuclear energy now prove the fundamental importance of the mass-energy relations. We may also hope that the entropy-information connection will, sooner or later, come into the foreground, and that we will discover where to use it to its full value.[14]

My second quotation will be from Eugene Wigner, in an analysis he gives of a book:

> There are very few complexes of phenomena which have been adequately described in terms of concepts developed in the earlier study of a much more restricted set of phenomena. Light turned out not to be a stream of particles which move according to the laws of mechanics; the study of electromagnetism did necessitate the introduction of the concept of fields. . . . These examples render it at least very likely that consciousness and life will be better understood on the basis of new concepts, yet to be developed, than on the basis of "our ordinary notions of physics and chemistry." If the emergence of such new concepts is what the author calls "vitalism," then the vitalists' point of view has a very good chance to prevail.

I really do believe that these two quotations from physicists have resonances consonant with the thinking devoted to finality, in its correspondence with low probability, by such great men and inspired writers as Bergson and Lecomte du Noüy.

NOTES

1. There is an essential connection between the so-called principle of special relativity that is the principle of equivalence of inertial frames, and the definition of time as a measurable magnitude. This was already true in Galileo's and Newton's days, since the concept of inertial reference frames is implied in the universal law $F = ma$. This law would need corrections in accelerated reference frames, and also if the Galileo-Newtonian time scale were replaced by an other one.

2. In fact Aristotle's "motion" comprised our kinematical motion as a special case of all conceivable changes in time. What was lacking in this perspective was the principle of inertia as a forerunner of the principle of special relativity.

3. According to Duhem, a new experimental fact never *imposes*, but rather *allows* and *suggests* a new postulate. And, according to Poincaré a postulate is equivalent to a definition. In the present case, Richelson's result allows Einstein's postulate that C is an absolute constant, and this, amounts to defining the length and line standards as physically equivalent.

4. The 11th International Conference of Weights and Measures legalized in 1960 the definition of the standard of length as the wave length of a particularly narrow emission line of Krypton 86.

5. Order is defined in such cases as any appropriate low populated subensemble of all permutations.

6. J. W. Gibbs, *The Collected Works of J. Willard Gibbs*, III (New York: Longmans, 1928), p. 150.

7. Reichenbach (not Grünbaum) contests this precise point, but I find his argument unconvincing.

8. In present days one should of course be alert to the implications of *PC*, or possibly *T* violations, in some elementary particle decays. To our cognizance, however, no truly convincing argument has been adduced for deducing so general a phenomenon as physical irreversibility from so rare and weak instances as *PC* violations.

9. This is denied by Popper but Watanabe, McLennan, Wu-Rivier, Penrose-Percival, and this writer, have produced arguments favoring the existence of the connection.

10. This is Renninger's negative form of the Einstein argument.

11. The Einstein-Podolsky-Rosen paradox proper was expressed in non-relativistic fashion. But a conception similar to ours can then be adduced: realism of the nonmetric Galileo-Newtonian space-time.

12. In telecommunications or machine computation, the significant sequence is rather $\Delta I_1 \geq \Delta N \geq \Delta I_2$.

13. Among philosophers, Descartes, Schopenhauer, Maine de Biran (however different their metaphysics) have strongly stressed the authenticity of willing awareness.

14. L. Brillouin, *Science and Information Theory*, 2nd ed. (New York: Academic Press, 1962), p. 294.

Pierre Lecomte du Noüy: Instrumentalist and Experimentalist

Charles Goillot

The two meetings held in 1967 (in May in Paris and in October at Notre Dame University) to commemorate the twentieth anniversary of Pierre Lecomte du Noüy's death have shown how alive his thought still remains. The conferences have also shown the importance of the problems he worked on, since scientists, philosophers, and theologians continue to come together to confront and discuss them.

Prior to being a philosopher, Pierre Lecomte du Noüy was an experimenter who strove to develop reliable apparatus, scientific methods of research, and reproducible results which led to those very conclusions he relied on to allow himself some philosophical reflection. Read only what he himself writes:

> A survey of the advancement of knowledge brings the realization that the progress of science has been achieved through the constant improvement of experimental techniques. This of course does not apply to the immense strides made by the speculative genius of men like Newton, Maxwell, and Einstein, who belong in a class apart.[1]

Even so, it has to be kept in mind that only experimental findings have given this genius its genuine importance.

He continues:

> The starting point of a discovery is often due to chance . . . but no matter how important, such chance is no more than a starting point and the future of the discovery depends on the *ability of the observer* to conceive and put into practice the methods which will enable him to dissect the phenomenon, to link it up with facts already known and to forge ahead.[2]

This is the value Lecomte du Noüy attributed to experience and to its interpretation. He makes his own this thought of Claude Bernard:

> Experimental criticism must be founded by creating rigorous methods of investigation and experimentation which will render the observations unquestionable and will eventually suppress the errors in the facts which are the root of errors in theory. The real promoter of scientific knowledge at present will be he who can bring some principle of simplification into analytical procedure, or improvement into the research apparatus. I am convinced that in the experimental sciences which are in evolution and particularly in those as complex as biology, *the discovery of a new instrument of observation or of experimentation renders much more service than many systematical or philosophical dissertations.*[3]

Suffice it to say that Lecomte du Noüy devoted much of himself to experimental training and that in fact he realized his plan:

1. He himself made the instruments which were indispensable for performing the scientific research he undertook. He built them and improved them thereafter. We will not even mention the many unavoidable tasks an experimenter is bound to perform, such as stabilizing the temperature of a bath. We shall instead devote ourselves only to describing a few essential instruments which still bear the name of Lecomte du Noüy and which now belong to the classical instrumentation of biophysical laboratories, and are included in basic textbooks.

2. As a result of critical analysis of the parameters involved in the actual measurement and of going back to the fundamental definition of the phenomenon, he was able to build an apparatus having scales which could be read in fundamental units. For example, to measure surface tension of liquids, he chose the only method which measures a force acting along a known length, because force-length ratio is the very definition of tension. Therefore his experimental results can never become obsolete even if widely accepted theories should finally appear to be wrong.

This is quite different from what had occurred in the newly developing biophysics. A look at figure 1, for example, will show that there were nineteen methods used to "measure" surface tension of liquids, as they were gathered by Allan Ferguson[4] out of more than 200 papers; nineteen methods which produced a huge mass of mere "observations" piled up in disorder, generally disconnected, showing discrepancies as great as 40 percent, and finally leading to inexplicable results.

3. Using these few but reliable apparatuses he built, he demonstrated that a systematic study, based on a sound working hypothesis, was capable of revealing an important number of phenomena; and by critically confronting the results, he opened up heretofore unthought of fields of research.

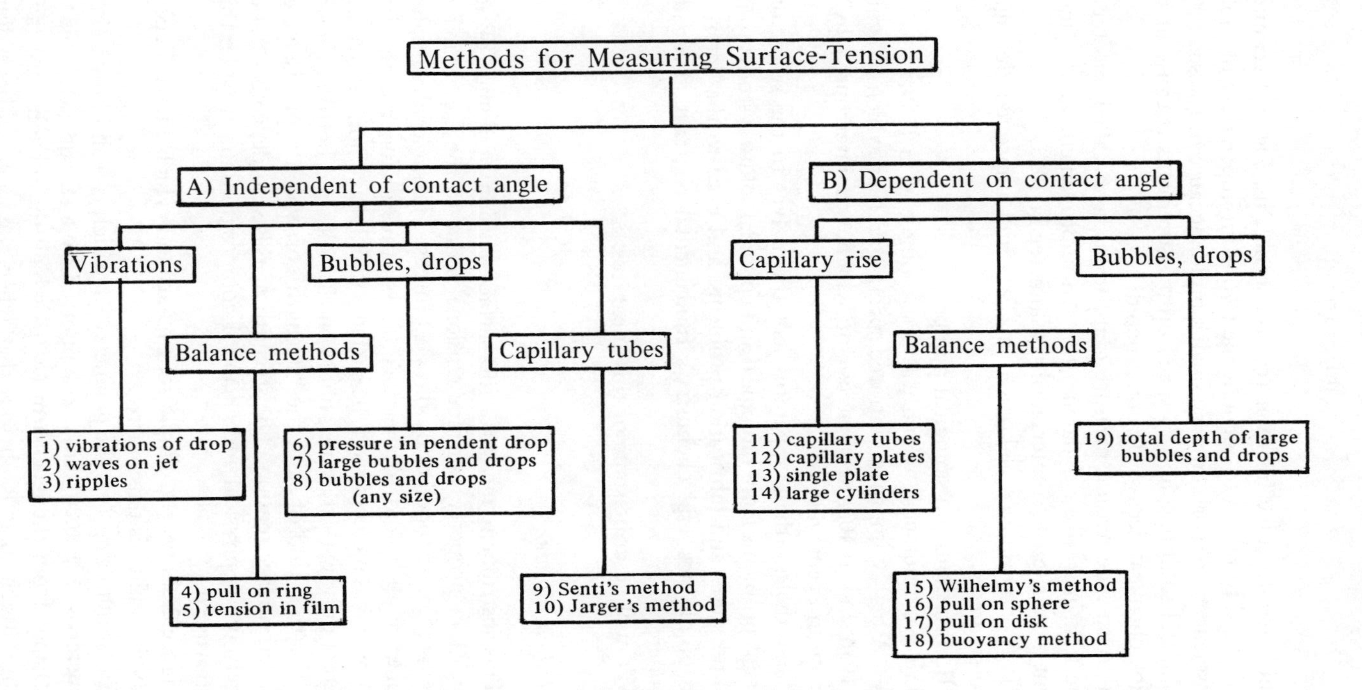

FIGURE 1. Various methods used to measure surface tension of liquids gathered by Allan Ferguson, 1914–1915.

APPARATUS

In order to solve his own research problems, Pierre Lecomte du Noüy had to take into account the two main imperatives of biological sciences; to work on very small samples because of the scarcity of biological solutions; and to speed up measurements because of their quick chemical decay. He constructed:

I. A surface tension measuring apparatus, a so-called tensiometer. The first instrument is dated 1919. An improvement of this instrument gave rise to an interfacial tensiometer in 1925.

II. A viscosimeter elaborated in 1923 and improved during the next six years.

III. Ingenious physiological pumps.

IV. An ionometer, which was first thought of in 1939.

V. A highly improved infrared spectrograph on which he worked in 1936 and 1940, the final use of which was interrupted by the European events of 1940.

Nevertheless, these apparatus lent themselves to the study of all sorts of problems from biological and colloidal solution study, to mineral oil study and lubrication problems, and even to the determination of physics' and chemistry's fundamental constant, Avogadro's number.

We will describe some of this apparatus.

The Tensiometer

This instrument measures the surface tension of liquids or solutions. This property (surface tension) is due to the fact that molecules are bound together by physical forces. The molecules inside the liquid are all bound to their neighbors (thus giving rise to viscosity), whereas molecules on the surface, which have none above them, interact only with their neighbors on the sides and bottom, thus giving rise to a kind of skin, the "strength" of which is precisely what is called surface tension. The strength of this skin depends of course largely on its physical and chemical composition and may be found by tearing it.

Figure 2 shows the simple scheme around which Pierre Lecomte du Noüy built his tensiometer.

A platinum ring (r), the geometry of which is well known (inside diameter = 4 ± 0,005 cm), is supported by a stirrup (s). This stirrup hangs from the lever-arm (a) to which the torsion wire (w) is rigidly attached. It is obvious that when the ring has been dipped into a liquid and lies on the surface layer of this liquid, a force must be applied to remove it from the surface. This force is provided by the torsion of wire (w), one end of which is secured in a limb (l),[5] which measures the rotation of this end, and thus the torque, because rotation and torque are accurately proportional. At the very instant when the ring is removed from the surface, the force applied by the film on the ring is balanced by the torsion of the wire. Thus the reading of the vernier at the instant the film breaks, gives the so-called surface tension of the liquid.

This "tearing off a ring" method was devised and used by Sondhaus in 1878 and Weinberg in 1902, in the form of a large copper

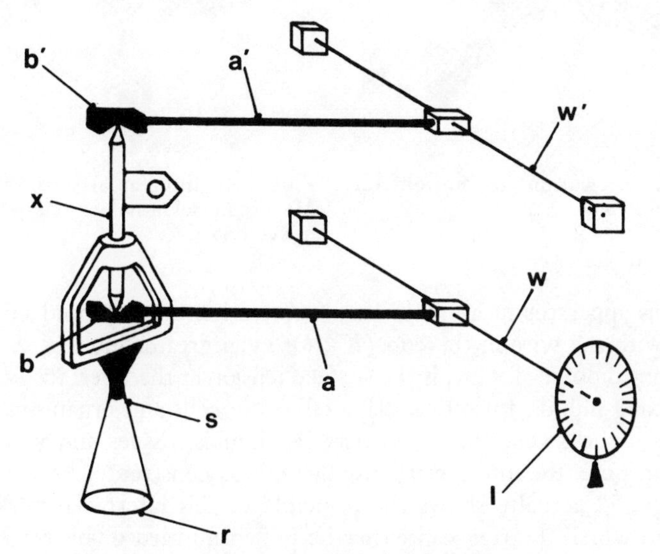

FIGURE 2. Principle of interfacial tensiometer.

ring hanging under the pan of a balance. This impractical method only led to slightly reliable results and was greatly improved by Pierre Lecomte du Noüy.

Following the scheme sketched above, Pierre Lecomte du Noüy designed and exhibited a tensiometer at the Congress of Philadelphia in 1919. This apparatus was then manufactured by the Central Scientific Co. (CENCO) in Chicago, and it was to be quickly improved to the point of being completely automatic by Rhône Poulenc Sté in France. Figure 3 shows this instrument, points of which are pointed by the diagram; (p) is a counterweight balancing the weight on the lever-arm (a), and in (K) is the cup which contains about 2 cc of liquid sample; the height of this cup is adjustable to set the zero position of the torsion wire (w).

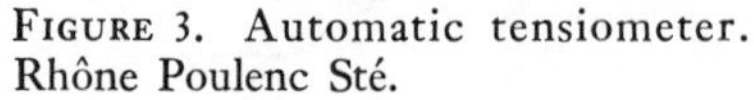

FIGURE 3. Automatic tensiometer. Rhône Poulenc Sté.

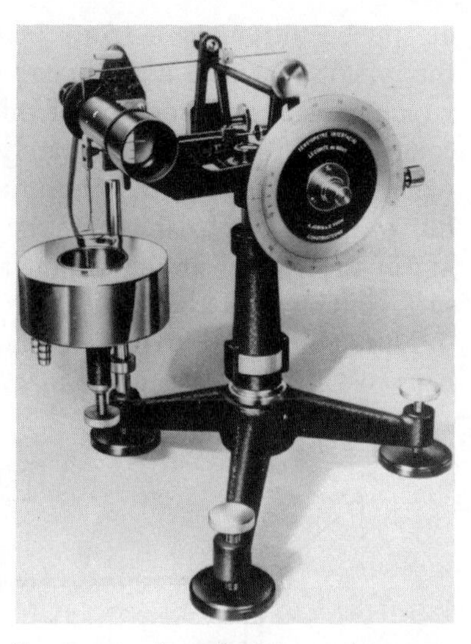

FIGURE 4. Interfacial tensiometer. Modern model manufactured by Jobin & Yvon Sté.

This apparatus measures the surface tension of the liquid in contact with the wire in the cup (*K*). Of even greater interest, from a biological point of view, is the surface tension at the interface of two unmixing liquids, for all capillary effects in cells and organisms take place at these very interfaces between liquids, tissues and gels; for this purpose, the interfacial tensiometer was constructed.[6]

Figure 2 actually shows the principle of this interfacial tensiometer in which the ring can either be pulled upward or be vertically pushed downward, according to whether the vernier is moved clockwise or counterclockwise, while its plane remains perfectly horizontal. To do so, the stirrup (*s*) [figure 2] is set on a vertical axis (*x*) inserted between two step-bearings (*b*) and (*b'*); a second torsion wire (*w'*) and its lever-arm (*a'*) make up an articulated parallelogram in which axis (*x*) remains vertical, thereby holding the ring horizontal. The interfacial tensiometer was manufactured in the United States by Cenco, in England by Cambridge Instruments, and in France by Jobin & Yvon Sté. Figure 4 shows an improved version of this last one, which includes light spot zero adjustment and thermostatic bath allowing the study of interfacial properties related to temperature. It is still in the Jobin & Yvon catalog and over the past thirty years about 300 models have been sold every year for use in both research and industrial laboratories in biophysics and organic chemistry.

Figure 5 shows the ring in the interface of two liquids, namely carbon tetrachloride on the bottom and water on top. It is shown exactly in the state of equilibrium when the holding strength of interface on the ring and the pulling force provided by the torsion of the wire are equal.

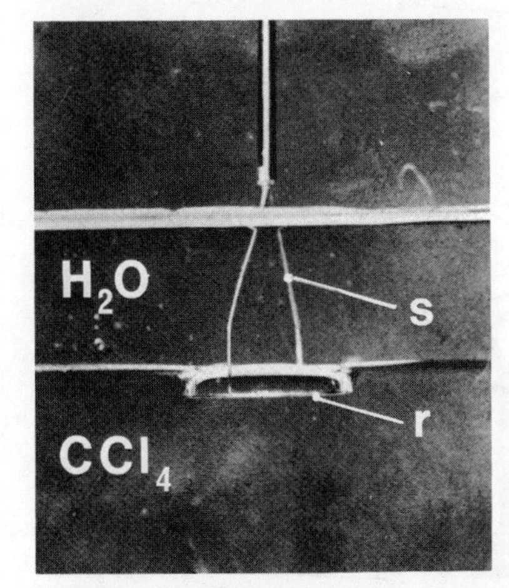

FIGURE 5. Ring of the interfacial tensiometer at interface H_2O/CCl_4.

More than 250,000 measurements have been made with these instruments at the Rockefeller Institute of New York and the Pasteur Institute in Paris, before conclusions could be reached about the significance of these measurements. The interfacial tensiometer has specifically allowed the study of surface properties between two biological solutions. We shall later discuss in detail some of the fundamental conclusions which have been reached with the help of these instruments.

The Viscosimeter

This apparatus directly uses the hydrodynamic properties of weakly bound molecules inside the liquids studied theoretically by Poiseuille. Figure 6 gives the diagram of this apparatus,[7] which is based on the principle of two axial cylinders with a liquid between them.

The outside cylinder, or cup (*K*) contains about 1 cc of liquid and rotates at a constant speed; the inside cylinder, or bob (*b*) is immersed completely in the liquid and is suspended by means of a thin phosphor-bronze ribbon (*w*) which carries a feeble current through a d'Arsonval coil (c) moving in the field of a horseshoe magnet (*H*). Under this coil is the mirror (*m*) and the hook from which the bob is suspended. The current passes through a milliam-

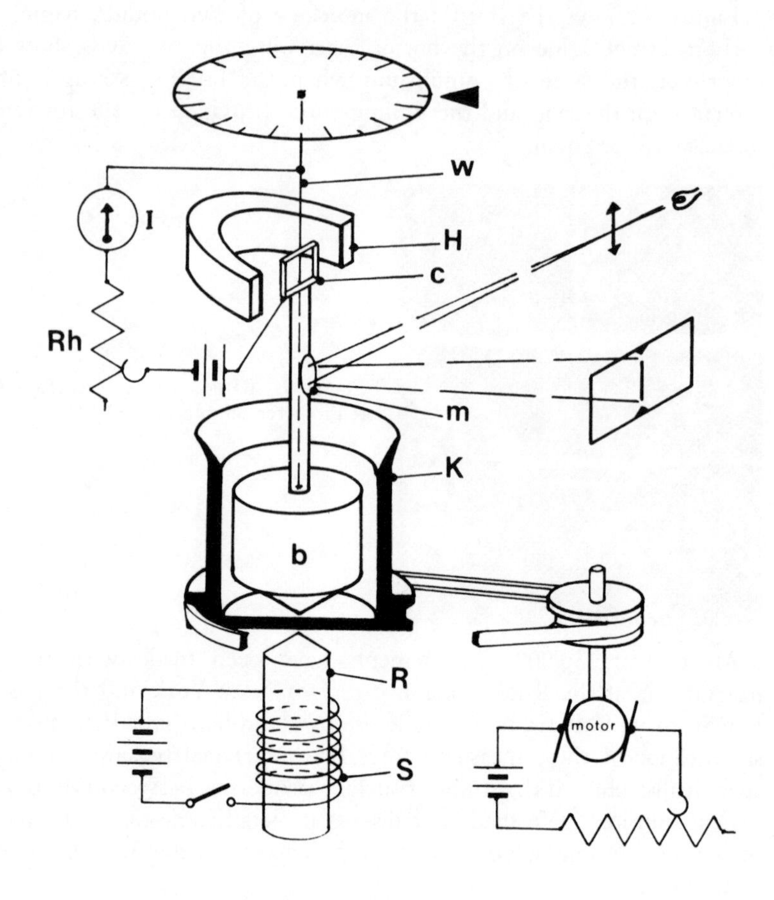

FIGURE 6. Scheme of viscosimeter with electromagnetic compensation and magnetic stabilization.

meter (*I*) and rheostats (*Rh*) so that it can be very accurately adjusted. The current sent through the coil counteracts the torque due to the friction set up by the motion of the liquid. As the spot is brought back to zero, a figure is read on the milliammeter which is proportional to the torque and thus to the viscosity of the liquid. By keeping the spot on the center of the scale, all further changes in the viscosity of the liquid under observation are quantitatively read on the milliammeter. An advantage of this electromagnetic control is that the readings are made when the suspension is *not* subjected to any torsion (zero position) and that the instrument damping may be adjusted as easily as in a galvanometer circuit.

Figure 7 shows a micro-viscosimeter as manufactured by Jobin &

FIGURE 7. Microviscosimeter with electromagnetic compensation.

Yvon Sté. Two other models were manufactured by Kaiser Wilhelm Institut in Germany, and Rhône Poulenc Sté in France.

This apparatus was decisively improved in 1939 by Pierre Lecomte du Noüy to prevent the inner bob from often going into a nutation motion. He received a patent for the magnetic arrangement he added beneath the cup as shown on figure 6: a soft iron rod (*R*) may be magnetized by the surrounding solenoid *(S)* so that the iron cone of the bob is magnetically centered upon the nib of this temporary magnet.

Suffice it to say that this viscosimeter was so sensitive that the viscosity of air (0.0001858 poises at 25° C) could be used to standardize the instrument which was able to give the sixth decimal point (0.000 001 poises).[8]

Finally, the instrument was made automatic and thus allowed the day and night recording of viscosity evolution according to time and temperature, without influence of operator skill.

Figure 8 is a drawing by Pierre Lecomte du Noüy, showing the bench arrangement for viscosity measurements as performed in his laboratory. We find again the viscosimeter with its numerous annexes: the electromagnetic compensation circuit, the thermic regulation circuit of the thermostatic bath, etc.

Pierre Lecomte du Noüy then dug deeper and deeper into the properties of biological solutions and grappled with their ionic prop-

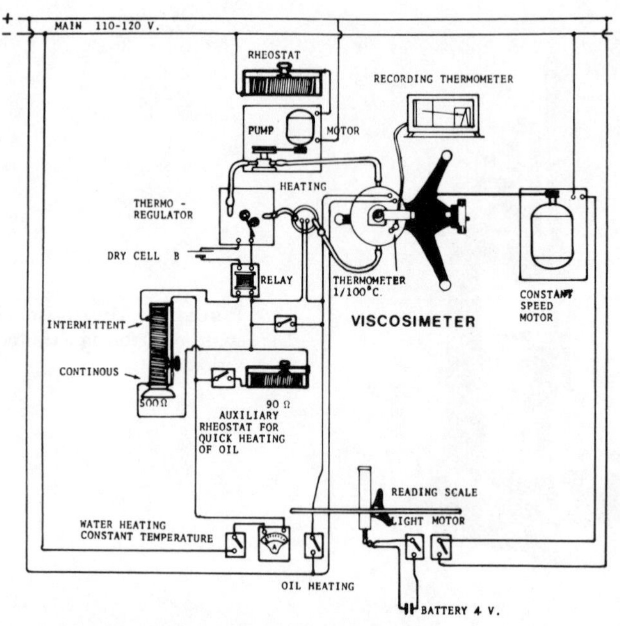

FIGURE 8. Bench arrangement for viscosimetry in laboratory of Pierre Lecomte du Noüy.

erties. Once more indispensable instruments were not in perfect order: the first hydrogen electrodes by Sorensen and Hildebrand required a long time to reach equilibrium. It frequently happened that potentiometer readings were still going up thirty minutes after the liquid had been in contact with the platinized electrode; it sometimes took more than forty minutes. Hasselbach showed that agitation by gas bubbling materially reduced the time necessary to reach equilibrium, but then the bubbles used made all measurements impossible and if paddling was stopped, instability again occurred.

Pierre Lecomte du Noüy overcame this difficulty with his "tilted rotating electrode," the principle of which is shown in figure 9.

A platinized platinum disk was mounted at the end of a shaft and tilted so that one half of its surface dipped into the liquid. The rotation of the shaft brought every part of the disk successively in contact with the liquid, while a thin layer of it clung to the platinum on its way out of the solution. As the whole was enclosed in such a way that hydrogen could be kept over the solution and as the bearing of the shaft was gas-tight, the liquid raised by the revolving disk was, so to speak, sandwiched between the atomic hydrogen absorbed on the platinum and the molecular hydrogen outside.

Figures 10 and 11 show a practical realization of the whole electrode by means of a syringe. The piston (*P*) has a platinum disk (*D*) welded at its end and contains a drop of quicksilver (*Hg*) to allow electrical potential to be carried out. The syringe contains a drop of the studied solution (*S*) and a captive hydrogen bubble

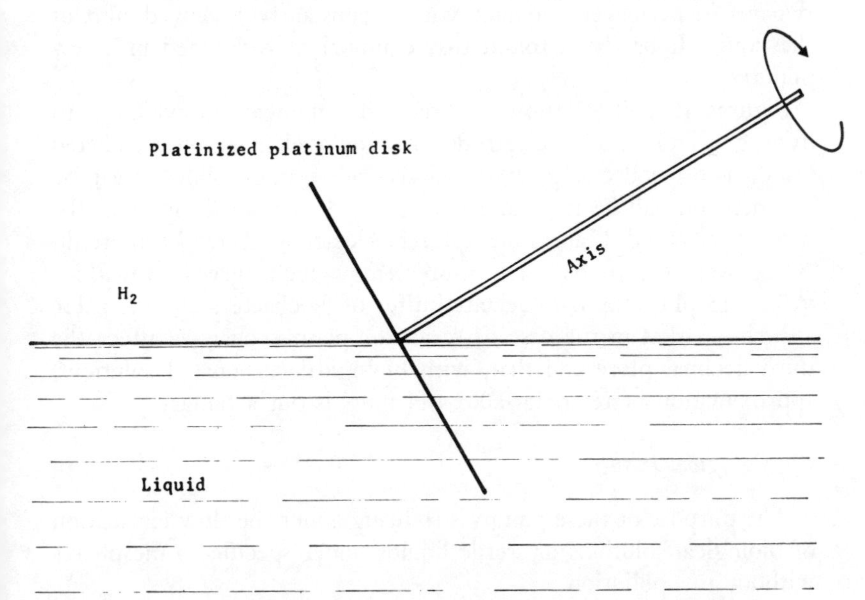

FIGURE 9. Principle of tilted and rotating platinum electrode.

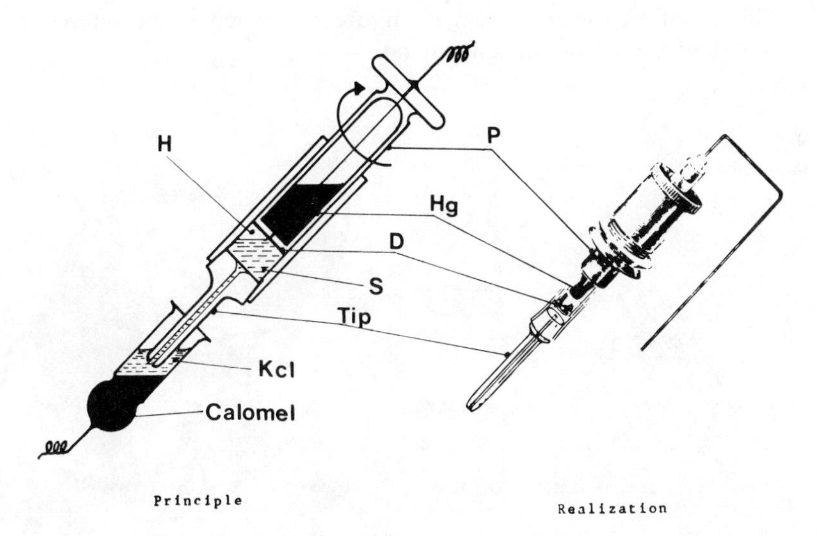

FIGURE 10. Principle of rotating syringe electrode.

FIGURE 11. Realization of rotating syringe electrode.

(*H*); the liquid junction to the calomel electrode is established through the capillary tip of the syringe. Instead of rotating the disk itself, the whole syringe is kept tilted and in rotation around its axis. The hydrogen bubble on top of the liquid, inside the syringe, remains in permanent contact with a constantly reviewed film of the liquid. It has been found that equilibrium is reached in a few minutes.

Figures 12 and 13 show a three-cell ionometer in which both hydrogen and calomel electrodes are enclosed in a water-jacketed stand. The metal casing, which holds each syringe, slides in a tube mounted on ball bearings and is connected to a small motor at the base of the stand. Using four different electrods, Pierre Lecomte du Noüy was able to obtain measurements which agreed to within: ± 0.00015 pH units with acetate buffer of Michaelis as solution. He also shows that in the case of serum or plasma, changes affect the third decimal place and that, with biological carbonated solutions, approximation better than 0.005 pH units is but a fantasy.

Physiological Pump

The purpose of these pumps is to bring about the slow circulation of biological solutions or sterile liquids under specified atmospheres without any pollution.

In 1919, Pierre Lecomte du Noüy patented in the U.S.A. and in France the pump shown in figure 14 which solved such a problem. This is still in use, as shown by figure 15, which is the reference 5-8951 of the 1966 Aminco catalog.

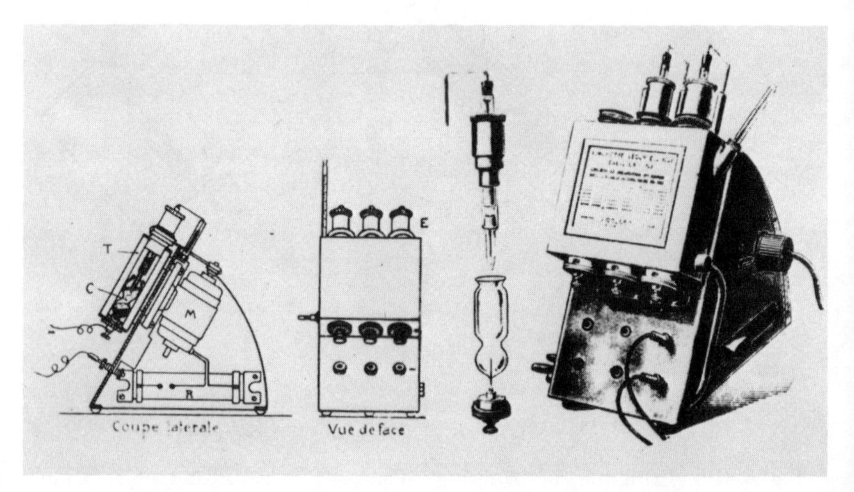

FIGURES 12–13. Three-cell ionometer. Rhône Poulenc Sté.

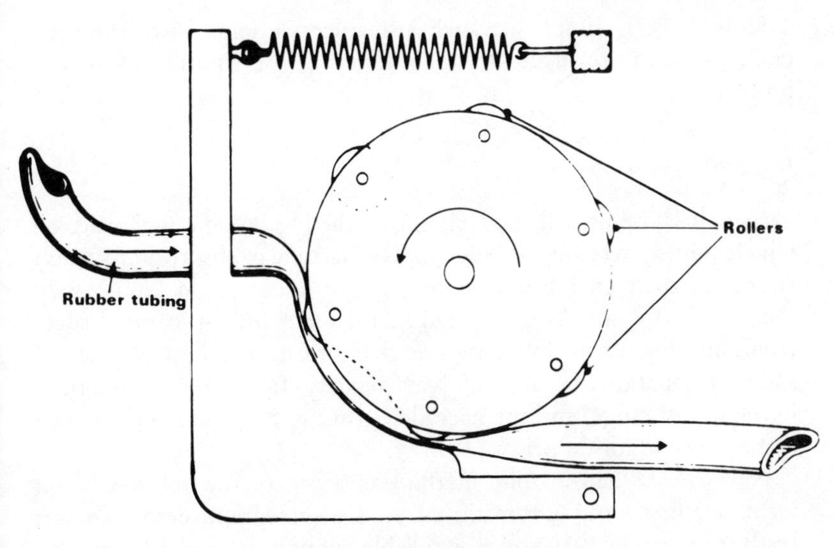

FIGURE 14. Physiological pump.

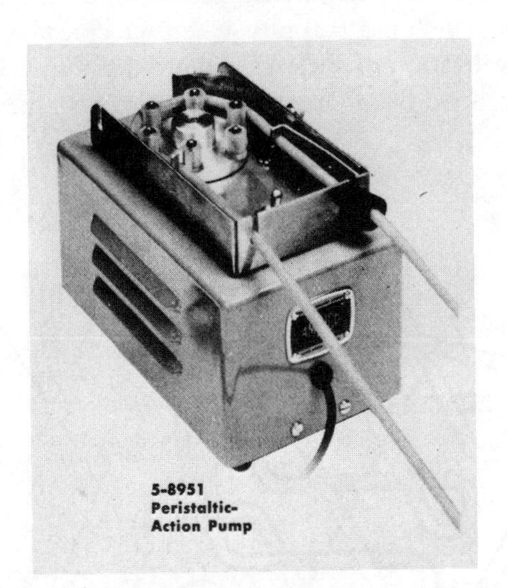

FIGURE 15. Peristaltic-action pump. Aminco (1966).

A disc which bears rollers squeezes the tube at two points, and when rolling on the tubing, causes a wave of liquid to be pushed forward.

Figure 16 shows an improvement of Lindbergh's "physiological pump," entirely made of glass and sealed permanently. Simply swinging around the axis, the sterile liquid is allowed to flow up to the top and then down to point (*A*), day and night, at will.

How cleverly simple are such instruments compared to the very complex literature of 1919 which aimed at the solution of such a problem!

Infrared Spectrometer

Now because he had the feeling that the big organic molecules on which he was working should display particular absorption spectra related to their chemical structure, Pierre Lecomte du Noüy, with the help of Jobin & Yvon, started in 1936 to build an infrared spectrograph. The main difficulty was to fabricate the optical parts of such an apparatus because of glass opacity for infrared radiations in the useful range, and the necessity of using crystalline optics such as a sodium chloride prism *(P)*.

Nevertheless and in spite of the skepticism of the Jobin & Yvon team, the first truly automatic infrared spectrograph ever seen was ready to work in 1939 and was reliable enough to give some results.

The events of 1940 were indeed to minimize the importance of this instrument, which is a real pity when we consider the beauty of the technical realization of Pierre Lecomte du Noüy. Only a decade later did infrared study of organic molecules become routine work

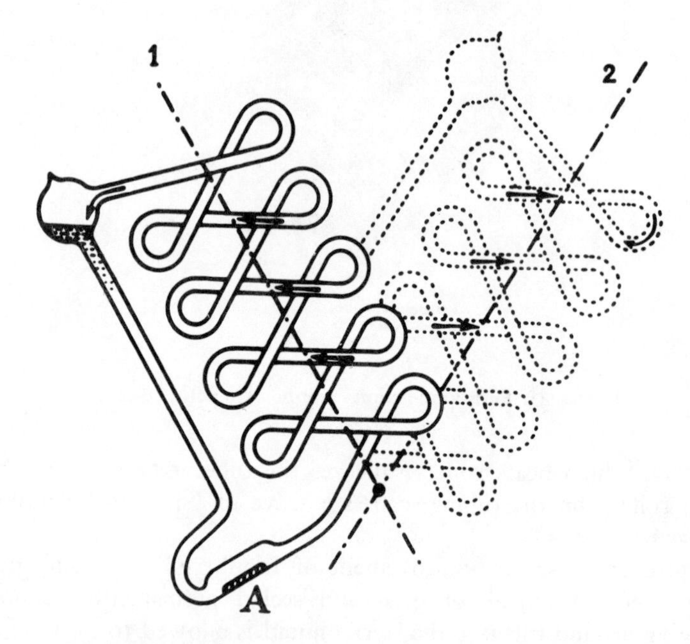

FIGURE 16. Improved Lindbergh's sterile serpentine pump.

among chemists, and gave rise to the astounding stride of synthesis in organic chemistry.

The detailed drawings of this apparatus were made by Pierre Lecomte du Noüy himself, and the development of parts by Jobin & Yvon workshops with which Pierre was in close contact.

Figure 17 shows two pictures of this beautiful instrument.

The light in (*S*) is a Nernst lamp, and the receiver (*R*) a photo-electric thermocouple cell, the signal of which is measured by a galvanometer and photographically registered on a drum (*D*). The apparatus is a three-beam spectrograph using a NaCl dispersion prism (*P*). It is fully programmed; the working spectral band may be chosen between 0.5 and 15 μ wave length, and the apparatus carries out automatically the sweeping of this band. Twice per second it successfully checks the electrical zero of the instrument, and the optical density of the studied solution. The wave lengths are automatically printed along the edge of the photographic recording paper, and both in and out slits are programmed according to the light beam energy density which falls on the entry slit.

How many laboratories would like, even today, to have such an instrument at their disposal!

RESULTS

It would require a very long list to enumerate all the definite scientific results obtained by Pierre Lecomte du Noüy. We shall cite

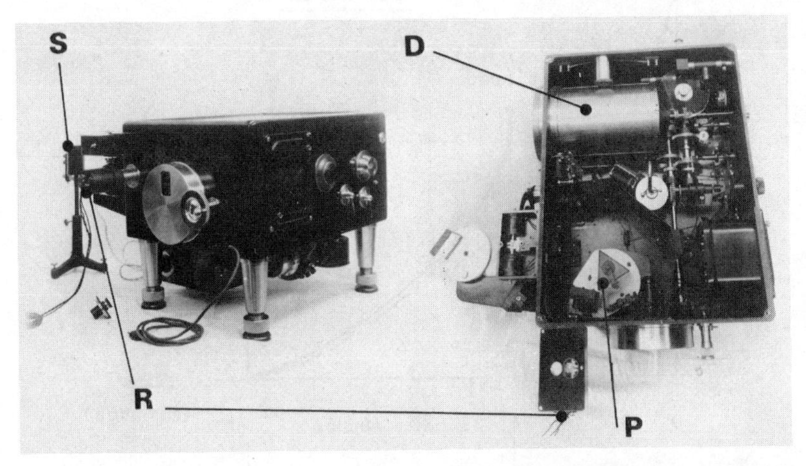

Side view Top view, cover removed

FIGURE 17. Three-beam, automatic infrared spectrograph. Manufactured by Jobin & Yvon Sté (1939).

only some of the most important ones, which decisively clarified the riddles propounded by colloidal solutions in those days.

TIME EVOLUTION OF COLLOIDAL SOLUTIONS

Owing to the very short time (less than ten seconds) required to perform one surface tension measurement by means of his tensiometer, Pierre Lecomte du Noüy was able systematically to follow the evolution with time of surface tension and hence to make obvious (as the graph shows in figure 18) that this surface tension: spontaneously dropped with time; and was partially and temporarily restored when the liquid was stirred.

Thus he pointed out that, for the same colloidal solution, there were indeed two main surface tensions to be considered—the initial one which he called the "dynamic value"; and the final one which he called "static value." (See figures 19 and 20).

These definitely clarified the problem of 40 percent discrepancies we have discussed and, of course, made obsolete a significant number of documents in which elapsed time had not been taken into account.

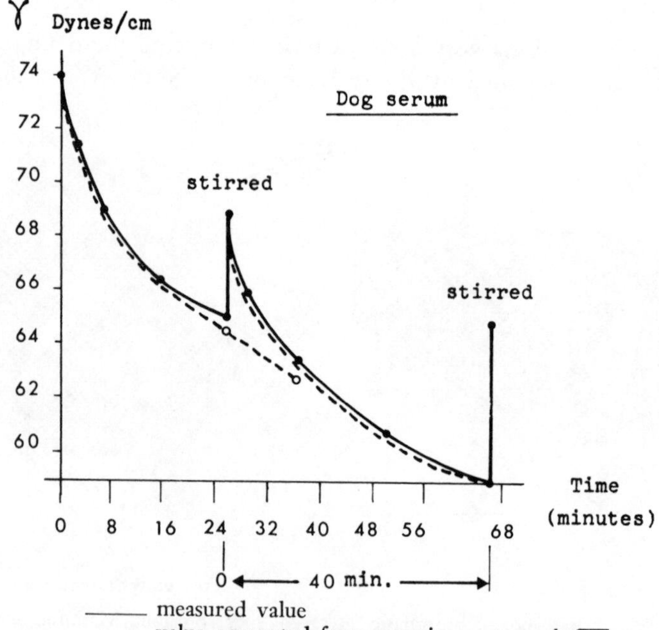

FIGURE 18. Dog serum solution. Time-drop of surface tension (γ) of serum diluted to 1/10. Influence of stirring of solution.

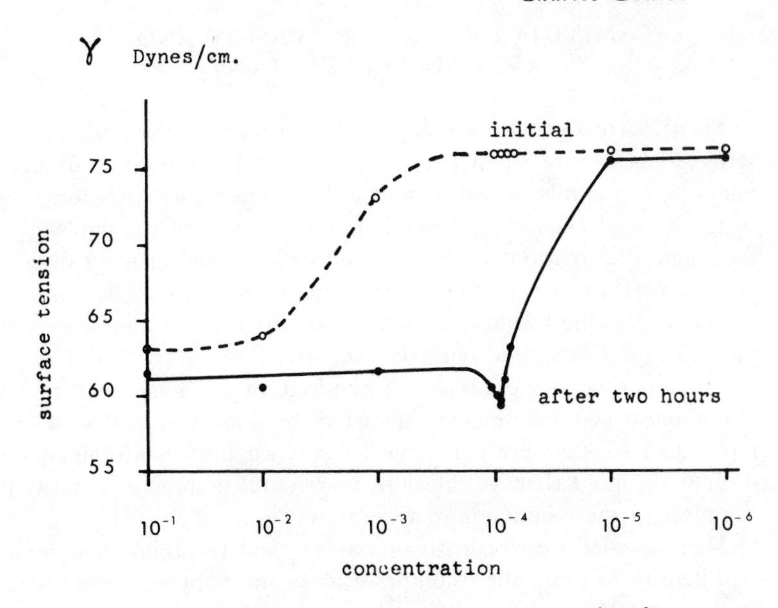

— — — dynamic value ——— static value

FIGURE 19. Rabbit serum solution. Surface tension of aqueous solution.

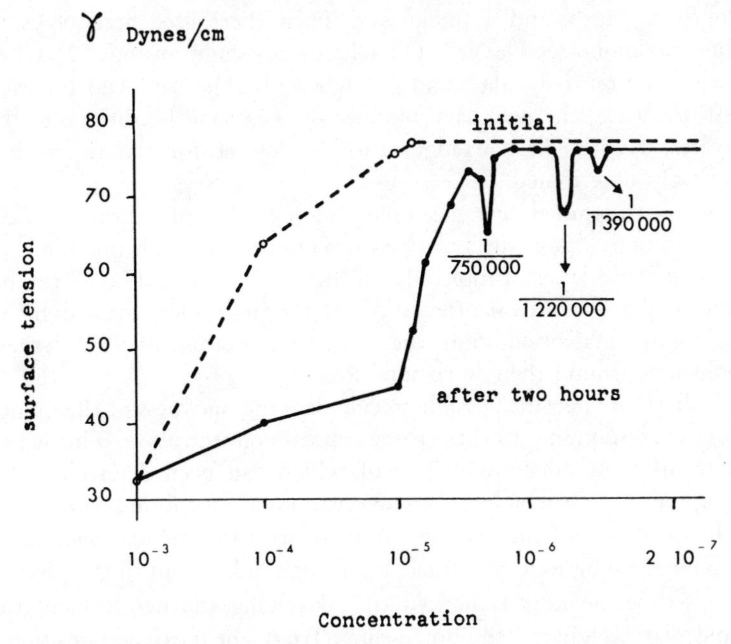

FIGURE 20. Sodium oleate solution. Plot of "static" and "dynamic" surface tension versus concentration of sodium oleate solution, showing the three minima in static surface tension.

MINIMA OF SURFACE TENSION AND AVOGADRO'S CONSTANT

Then Pierre Lecomte du Noüy plotted static surface tension of serum solution versus concentration; instead of a smooth continuous curve he found minima on it at well determined concentrations, the value of which, however, depended upon the size of the cup which contained the solution! In order to understand this amazing discovery, he performed the same experiments on sodium oleate solution (a simpler colloid solution) and discovered three minima[9] (see figure 20) which shifted with the cup size.

The experiments were repeated hundreds of times and they represent about 8800 measurements of surface tension. As a matter of fact more than 10,000 were performed, for when perfect samples were dealt with, more than 44 dilutions were studied so as to determine the place of the minima more accurately.

Then he tried to explain this discovery and the following model will help us to grasp the hypothesis he set up. Suppose we let seeds float on quicksilver with a limited surface. If these seeds are spherical, the number of them necessary to constitute a "monoseed" layer on the surface is perfectly determined. The problem is quite different with three dimensional seeds (pumpkin seeds, i.e., which have a length, a width, and a thickness). Then there are three ways to obtain a "mono-seed layer": (1) when seeds stand on top; (2) when seeds stand on their edge; and (3) when seeds lie flat. And for each of these ways, the necessary number of seeds will be different (the greatest for the first arrangement, the lowest for the third) but perfectly determined.

So he concluded that a simple and logical explanation of these three minima lies in the polarized organization at each minimum of a single mono-layer of molecules in the surface layer. The first minimum is due to vertical orientation of the molecules; the second to their horizontal orientation; the third to a rotation of 90° of these molecules around their horizontal axis.

It therefore became possible to calculate the thickness of the monolayer corresponding to these three critical concentration. This led to three different dimensions (one of which had been determined by Langmuir) from which the volume is known (see figure 21).

Proceeding still further with this investigation, which was taking him far from biology of course, but which linked him to the physics of his time, he now remarked that, knowing the density and the molecular weight of sodium oleate (from chemical composition), the volume of the molecule and the concentration where minima

CH = CH
CH_2 CH_2
CH_2 CH_2
CH_2 CH_2
CH_2 CH_2
CH_2 CH_2
CH_2 CH_2
CH_2 CH_2
CH_2 C-O
H O
Na

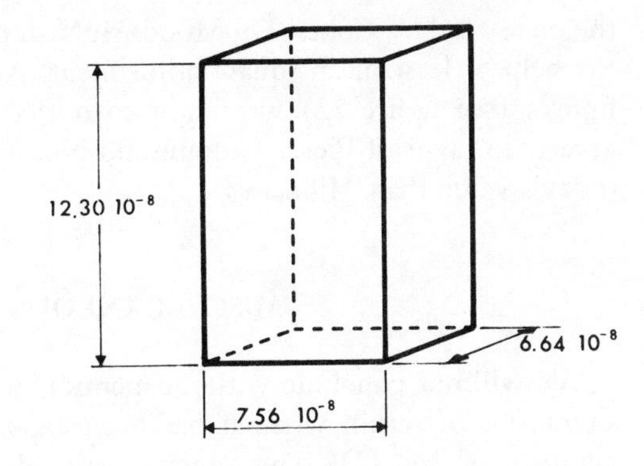

lengths in cm.

FIGURE 21. Sodium oleate molecule. Developed formula and apparent volume deduced from surface tension minima measurements.

(mono-layers of known surface) occur, he was able to compute by mere geometric considerations one of the most fundamental constants of physics and chemistry, Avogadro's number (N), he was led to the value:[10]

$$N = (6.004 \pm 0.009)\ 10^{23} \text{ molecules/mole}$$

He compared his own value to the one given by Millikan (Nobel prize) from measurements of electron charge: $(6.062 \pm 0.006)\ 10^{23}$ and humbly said:

> Nevertheless it is not in our intention to claim for this value a greater accuracy than that of the figure published by Millikan as a result of his admirable and exact measurement of the charge of a single electron. The difference of 1 per cent between the two figures indicates there is probably another cause of error in our experiments, or that we have underestimated those that we mention. The interest of the figure obtained by the above described method lies chiefly in the fact that it confirms our hypothesis concerning the monolayers and that it is derived from the only method based on the very definition of the Avogadro Constant. To our knowledge it is the most direct and simple method yet devised.

It is of great interest to compare the figure obtained by Pierre Lecomte du Noüy with his "surface tension method" not only to the figures of his time, but also of those of our own time. Today, after numerous experiments which might be the topic of a whole colloquium, we admit as the most reliable value of Avogadro's constant the one stated by Cohen-Du Mond-McNish of the NBS in 1961 with the help of least mean square adjustment. And if we compare these figures (see figure 22) we ought to notice that the comparison is greatly in favor of Pierre Lecomte du Noüy whose value is closer to today's value than Millikan's.

MISCELLANEOUS

We will not conclude without mentioning the twelve irreversible alterations of serum when it has been exposed to the critical temperature of 56° C.[11] The "antagonistic phenomenon" in colloidal solutions where a colloid can momentarily cancel surface tension properties of another, thus allowing analytical methods able to detect as little as 1.10^{-8} g (1/1000000000) of protein in a solution.[12]

We shall refer those who wish to know more about this to Pierre Lecomte du Noüy's own works.

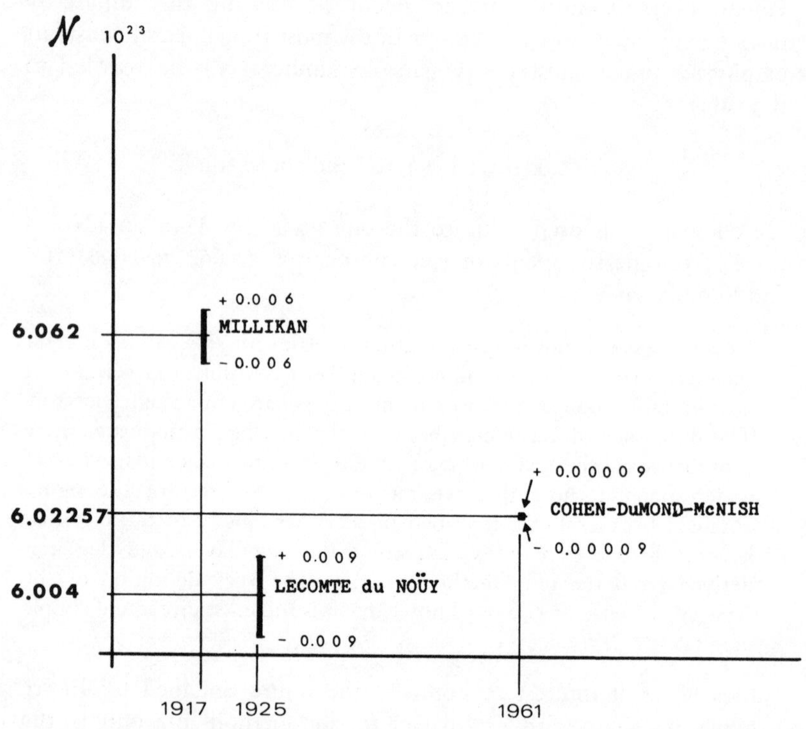

FIGURE 22. Comparison of Avogadro's numbers computed by Millikan, from electron charge measurements (1917); Lecomte du Noüy, from surface tension of sodium oleate solutions (1925); Cohen-DuMond-McNish, least square adjustment NBS value (1961).

To conclude, and looking back as we can now do, we must admit that, in his time, Pierre Lecomte du Noüy brought about a complete renewal of many scientific controversies. His tensiometer, for example, crowned a discussion which ended only in 1930 when the two Freuds finally wrote the whole theory of the instrument and pointed out that it could be used as absolutely reliable. Du Noüy built instruments but he also, with these instruments, expounded absolutely unsuspected properties. At last, although working on complex solutions, he was able to link the results he had found to fundamental phenomena of the physical domain. How could we avoid the conclusion that his mind was essentially that of an experimentalist when we hear him saying:

> It is wise to grant theories, especially those which do not proceed from precise experimental facts, only an importance suited to the amount of progress they have brought—or are thought liable to bring.

He was also a philosopher. To be convinced, one must only notice that he philosophized after acting and consider how his philosophical works follow the development of his scientific ones.

NOTES

1. Pierre Lecomte du Noüy, *Surface Equilibria of Biological and Organic Colloids* (New York: Reinhold, 1926), Introduction.
2. *Ibid.*
3. Claude Bernard, *Introduction à l'Etude de la Médecine Expérimentale*, 4th ed. (Paris: Flammarion, 1920), pp. 272–273.
4. Allan Ferguson, "Capillary Constants and Their Measurements," *Science Progress* 9 (1914–1915), p. 428.
5. Pierre Lecomte du Noüy, *J. of Gen. Physiol.* 1 (1918–1919), p. 521.
6. Pierre Lecomte du Noüy, *J. of Gen. Physiol.* 7 (1925), p. 625.
7. Pierre Lecomte du Noüy, *J. of Gen. Physiol.* 5 (1923), p. 629.
8. It has to be said that a difference of 0.01°C between 20 and 30°C introduces a variation of two units in the fifth decimal place.
9. Pierre Lecomte du Noüy, *Phil. Mag.* 48 (1926), p. 664–672.
10. The whole detailed calculation is given in *Surface Equilibria of Biological and Organic Colloids*.
11. Pierre Lecomte du Noüy, *Studies in Biophysics: The Critical Temperature of Serum* (New York: Reinhold, 1945).
12. Pierre Lecomte du Noüy, "A Highly Sensitive Method for Detecting Proteins and Colloidal Particles in a Solution," *Science* 61 (1925), p. 472.

The Scientific Career of Pierre Lecomte du Noüy

Karl K. Darrow

I will begin by telling you something about the preparation of this speech. When Madame Lecomte du Noüy invited me to address you on the scientific work of her husband, I was about to start on rather extensive travels. With me I took the *Exposé des Travaux de Pierre Lecomte du Noüy* published in 1937. This was an excellent choice; I do not think I could have made another nearly so good. It is perhaps a disadvantage that I cannot speak of the continuation, development, and sequels of his work performed by others in subsequent years. This, however, I could not have done even if I had settled next door to the best library in the world, for it is to be presumed that these continuations and sequels are to be found in the biological and biochemical literature, and I have no competence to search or understand this. By the way, any unintended gallicisms in my talk must be ascribed to the fact that I read Pierre's work in French.

One may be tempted to say that Pierre Lecomte du Noüy was a scientist before he was a philosopher. I prefer to say that he was always a philosopher, and one who for twenty-one or twenty-two years was also an experimental scientist of diligence and distinction. On experimental science, both of his own making and of the making of others, his philosophy was solidly based. I have often read that some of the Greek philosophers believed that truth could be attained without making any observations or experiments. If this is so (and I do not say of my own knowledge that it is), they were at the opposite pole from Pierre.

How did Pierre happen to become an experimental scientist? This is a remarkable story, which he has himself related. Neither heredity nor environment played any discernible part. Born into an artistic and literary atmosphere, he developed a great variety of interests; but in the midst of these he found himself possessed by so strong a taste for scientific questions that he entered a yearlong course in mathematics with a special objective of mechanics. To this year of mathematico-mechanical studies he credits his skill as a designer and maker of instruments. As he mentioned, this was the period when the discovery of radioactivity (made, I remind you, at Paris) was kindling a general interest in physics. I remember that in my student days a decade later, the fascination of radioactivity was still intense. I also remember that some of my classmates considered mechanics a great bore.

After this year Pierre turned away again from science, and he was engaged in juridical, literary, and philosophical studies when one of those apparently trivial events took place such as sometimes determine the course of a lifetime. One evening he went to a banquet of the newly formed Société de Chimie Physique. This dinner steered Pierre into science. There he met the British scientist Sir William Ramsay, who invited him to visit his laboratory in England. Pierre went, and there he met another British scientist, William Bayliss. The conversations that he had with these gentlemen (and how one would love to have a tape recording of them!) inspired him to return to science, and to attend the courses given by great scientists at the University of Paris. Then came the First World War, and Pierre became a soldier on its second day. That banquet of the Société de Chimie Physique may well have saved Pierre not only for science but for this world. He would probably have been one of the million or more young Frenchmen who are remembered now only by the aging and the aged, had not his scientific talents given him a special status.

This special status did not come to Pierre immediately. He was mobilized at Compiègne, and in that very city Alexis Carrel of the Rockefeller Institute (now Rockefeller University) of New York was organizing a hospital. It was in 1915 that Pierre, having made Carrel's acquaintance, undertook the study of the healing or cicatrization of wounds—a study for which the circumstances of the time and the place gave material in abundance. This was the debut of Pierre into science, and it brought him his doctorate at the University of Paris and two trips to the United States, all in wartime. Now we will look at this, the first of what Pierre regarded as his three major enterprises.

The procedure was to compare the area of a wound at four-day intervals up to its healing. The technique was to apply a leaf of some transparent material to the wound, trace the contour of the wound upon it, and measure with a planimeter the area enclosed by the contour. The observations were plotted as area against time; mathematical formulas were found that gave curves fitting the data. There were two or more of these formulas; it would be pointless to reproduce them here or ask whether one was right and the others wrong, for they were purely empirical and descriptive. With them it was possible to extrapolate, and with success, to the time of healing.

Strikingly shown by these experiments was the dependence of the rate of healing upon the age of the person. I quote: "If a child of ten years heals a wound of 20 cm in 20 days, a man of 20 will heal it in 31 days, a man of 30 in 41 days, a man of 40 in 55, a man of 50 in 78 and a man of 60 probably in 100 days. . . ." The data for ages 20, 30, and 40 being much more abundant than those for the other ages (and I think we can all readily guess the reason why), the figures for these ages are the most accurate. On such facts as these was founded Pierre's book *Biological Time.*

An interesting by-product of these investigations was the following. Before each measurement it was ascertained whether the wound was bacteriologically sterile. Usually it was, but sometimes not; and in the latter case the cicatrization of the wound was slowed down, but only to rebound when the wound was again made sterile. In fact, the rate of healing actually speeded up after the sterilization, as though nature were making up for lost time. These observations allowed evaluations of the efficacies of various antiseptics, with results which perhaps were not always pleasing to the manufacturers of them.

The war over, the full import of Pierre's meeting with Carrel became apparent. Through Carrel's recommendation he was invited to come to the Rockefeller Institute, which for eight years was the theater of his work. Here he could devote himself to problems of less urgency, though not of less basic importance, than the healing of war-caused wounds. Serum was the principal though not the only substance with which he worked, and the problem of immunization gave him his strongest incentives to work on it. I mention here his eloquent statement of a contrast between research on biological problems and research on physical problems. In the latter, the physicist tries to proceed by simplifying the situation until the phenomena become intelligible. But in biology, if one starts to simplify the situation one may find that before he gets very far he has simplified out of existence the phenomena which he set out to study!

Just in passing I mention that one of the first, perhaps the very first undertaking of Pierre at the Rockefeller Institute was to measure the sensitivity of the human eye; but as he relegates his account of this to the end of the *Exposé* and as I suspect that his results have been superseded, I go right on to the second of his great enterprises which was a study of surface tension. Here his ability as a designer and maker of instruments came to the fore. His object was to develop a method of measurement which should require only small quantities of liquid, and with which measurements could be made in rapid succession. The preferred method is called in French *arrachement*: a metal ring is dipped into the liquid, and drawn up and out. It draws a thin film after it, and the quantity which is measured is the force at the moment when the film is ruptured. The "tensiometers" which Pierre devised were later made in mass production by several firms. With one form of the instrument it is possible to measure the "interfacial tension" of the surface that separates two nonmiscible liquids.

Now it is time to turn to the results, but I digress for a moment to say that it was in Pierre's "Rockefeller" period, and probably at an early date in it, that I first saw and heard him speak. I vividly remember that he spoke with a strong French accent. The only reason I have for mentioning this is that in later years he spoke English perfectly, and I mean *perfectly*. This is encouraging, for it shows that it is possible for one to learn even in maturity to speak a foreign language without accent. Of course this ability may be confined to people as bright as Pierre.

Now for the experiments. Highly diluted serum was shaken up vigorously to homogenize it and establish what was taken as the normal point of departure, or initial condition. The surface tension was then measured at intervals of a few minutes, the liquid having come to rest; and it was found to diminish steadily with the lapse of time, sometimes approaching and in some cases reaching a stable new value which could be as low as two-thirds of the initial value. We do not know for sure what the truly initial value should have been, since the tensiometer cannot work with infinite speed to measure the tension as soon as the liquid comes to rest. The limiting or ultimate value of the surface tension was found to depend upon the dilution of the serum, presenting a minimum at a certain value of dilution. On the contrary, it was a function of the ratio of the surface S to the volume V of the liquid. On this fact Pierre based his theory that the diminution of the surface tension is due to the sudden rush of the molecules of the serum to the surface of the diluting liquid, there to form a surface layer. If the ratio of the surface S to the volume V has just the right value for the dilution in question, the layer will be monomolecular. It is a further part of Pierre's theory that when for a given ratio of S to V the dilution is such that the surface tension is at its minimum, that is just when the layer is monomolecular.

If this is so, and if we know from other sources the dimensions of the molecule, then we know how many molecules are laid down upon a unit area of the surface, therefore how many molecules there are in the surface layer altogether. But this is the same as the number of molecules that were originally dispersed like a gas through the liquid, and dividing this number by the volume, we get the number of molecules of the solute per unit volume of the liquid. But this number is calculable from the known dilution of the serum and from what chemists call Avogadro's number. Thus Pierre's theory can be checked, provided that one makes the experiments not upon serum of which the dimensions of the molecules are (or were) unknown, but on solutions of some substance of which the molecular dimensions *are* known from other sources. Such a substance is the one which in French is called *oléate de soude*—I translate this as sodium oleate, without knowing that I am right. The dimensions of the molecules of this substance had been determined by Langmuir. Taking the best value of Avogadro's number then extant—the one due to R. A. Millikan—Pierre found that his theory of the monomolecular surface layer was well checked. We can say exactly the same thing in another way: Pierre, assuming the correctness of his theory, calculated a value for Avogadro's number and this agreed with Millikan's value. This latter is the way Pierre expresses it. Experiments of the same kind were performed upon ovalbumine with similar but not so accurate results.

Now we look at another remarkable phenomenon. Sodium oleate was added, in small amounts, to undiluted serum taken from a dog. The mixture was shaken up, and after it was again at rest the evolution of the surface tension was followed by measurements which were taken as rapidly as possible, a minute or less apart. At first the surface tension diminished very rapidly, just as I have told you that it does in the case of diluted sodium oleate. But in a minute or so it started to rise swiftly, and soon it regained its initial value. This is as if the molecules of sodium oleate had started their rush to the surface just as they would in pure water, and then the molecules of the serum had grabbed them and pulled them back and held onto them. Such is in essence Pierre's conception, though the words he uses are somewhat less colloquial. In his terminology the sodium oleate is called a surface active substance, what the molecules of the serum do to its molecules is called neutralization, and is said to result from antagonizing forces. There are other surface active substances, and there are other liquids than serum that contain neutralizing molecules. In an example cited by Pierre, the liquid is a colloidal suspension of gold and the surface active substance is again sodium oleate.

Remarkable and potentially helpful is the fact that by this phenomenon, a quantity of a neutralizing substance (proteins, for example) that is too small to produce any detectable effect on the surface tension of water is nevertheless large enough to produce a detectable effect by neutralizing sodium oleate. The phenomenon thus offers an exceptionally delicate method (perhaps unique) for detecting certain important substances.

At this point in his *Exposé* Pierre gets closer to his ultimate objective, the study of immunity; and at this point also, I get out of my depth. It is clear that he continues with the procedure that I have just described: he shakes up a liquid to homogenize it, and then after it comes to rest he studies the evolution of its surface tension with time. The liquids are samples of serum taken from rabbits, dogs, and chickens, dissolved in the isotonic solution of NaCl. Always the tension diminishes from its initial value to a relatively stable value. I use Pierre's own word *time-drop* to signify the difference between the initial value and the relatively stable value. Sometimes Pierre introduced antigens into the animals; these I call "immunized" animals. Other animals he left nonimmunized to serve as controls. In his latest work, he made measurements on the serum of one and the same animal both before and after immunization. The general result is that if the time-drop before immunization is low, it becomes considerably higher after immunization; the change is of the order of two to one. This is a strong indication that a high value of time-drop is a sign of the presence of antibodies. But also there were animals for which the time-drop was high initially, and was not enhanced by the introduction of antigens. Pierre accounts for these by postulating that in these animals, antibodies and immunity were already present. It is interesting that studies on human beings showed that adults always have high time-drops; low ones are found in young children only.

Now the year advances to 1927, the scene shifts from the Rockefeller Institute in New York to the Institut Pasteur in Paris where Pierre went that year, and we arrive at the third of what Pierre regarded as his three major enterprises. Here I am again out of my depth, but I must attempt to give you some background. There are two "critical points" or "critical temperatures" of serum, at each of which what I can only call biological changes occur. These are near 56° and near 66°, respectively. The former is said to correspond to the destruction of alexine, the latter to the destruction of the sensibilizer (*sensibisatrice*). These words convey nothing to me, but I take consolation from the fact that Pierre makes it fairly clear that he had no very flattering opinion of the concepts which they presumably embody. Pierre's ambition was to find physical and chemical attributes of serum which show a dramatic or at least a marked alteration at one or the other or both of these two temperatures. In this he was very successful, for he enumerates no fewer than twelve such attributes which he found. Of this wide panorama we will look only at some of the high spots.

Viscosity is the first of these attributes. In his measurements of this quantity, Pierre's talent as a designer and maker of instruments came again to the fore. The method that he used was not a new one, but the apparatus that he built was so proportioned that it could make successive measurements quickly and could make them on a single cc of liquid. With this apparatus he discovered that something dramatic does indeed occur between 56° and 59°: a sharp minimum in the viscosity. As the temperature is raised toward 56°, the viscosity at first falls off in proportion to that of water; then it turns and rises. For the ascending portion of the curve there is a theoretical formula. This formula, credited to Kunitz, expresses the theory that the rise in viscosity is due to a diminution in the volume of water and an increase in the volume of the particles swimming about in the water. But how can this be, since neither the amount of water nor the quantity of serum has been changed? Pierre's conception is that as the temperature rises, a steadily increasing fraction of the water becomes affixed to the particles and thus loses its fluid character. This is what Pierre calls hydration of the serum.

Next, Pierre examined the rotation of the plane of polarization of a beam of polarized light traversing the serum. This quantity or attribute, sensibly constant until the temperature of 52° is reached, increases steadily after this "threshold" temperature is passed. Well, a curve which is horizontal up to 52° and then bends upward is perhaps not so dramatic as a curve with a minimum in it; but it is significant nonetheless, and here accordingly is a second attribute of serum which shows that something happens when the temperature gets into the fifties. This something in Pierre's theory is hydration. But why should hydration, conceived as an adhesion of water molecules to serum molecules, wait until such a temperature as 52° to begin? Pierre's conception, as I read it, is that the water molecules burst their way into the loose structure of the protein molecules and take up their positions inside this structure, the protein molecules swelling to make room for them. The significance of the critical temperature is, then, that it is the temperature at which the energy of thermal agitation of the water molecules enables them to burst into the serum molecules and entrap themselves therein.

I have spoken of two of the twelve attributes which form, so to speak, the symptoms of the syndrome attending the critical temperature of serum, and now I will allude to just two more. One of these is the sidewise scattering of a beam of light passing through serum—what used to be called and is still sometimes called the "Tyndall effect." Using a powerful adjective rarely found in scientific writing, Pierre says that this scattering increases "brutally" from 55° onward. By a theoretical formula Pierre calculated that the molecules of serum swelled up to fivefold their original value. The other attribute is the absorption spectrum of serum in the near ultraviolet. This is characterized by a tall maximum at about 2,750° and a deep minimum at about 2,500°. These two features are remarkably little dependent on, in fact almost independent of, the nature and the temperature of the serum. Yet there are minor but definite alterations as the upper of the two critical temperatures is approached and attained, and this attribute seems to be a particular sign of the change at 66°.

In a sort of epilogue to his *Exposé*, Pierre tells us that when to an immunized serum is added some of its *specific* antigen, the viscosity rises quickly—in a time of the order of minutes—to a sharp high maximum, and then as quickly recedes to something near its original value. Nothing of the sort occurs when some of a *different* antigen is added. The same thing happens when diphtheria antitoxin is added to diphtheria toxin: the time required for the viscosity to climb to its maximum is a sensitive function of the ratio of antitoxin to toxin, and could be used to determine this ratio.

You must not conclude that I have told you everything that Pierre Lecomte du Noüy did in those twenty-two years of devotion to experimental science. Much has been left out; for example, his observations on the changes in serum that is kept at a constant warm temperature for a long time. These observations are numerous and not readily generalized. Suffice it to say that heating for an appreciable time at temperatures near the critical ones is likely to bring about irreversible changes.

There is an old and very stale joke about the university dean or president who evaluates the members of his faculty by counting or weighing their publications. This joke almost frightens me away from giving the statistic of Pierre's papers. But I do not allow myself to be frightened, for it is a valuable and a meritorious fact that in these years Pierre published 112 papers and 6 books. Indeed this was a notable harvest from twenty-two years of planting and cultivating and reaping. I only hope that I have succeeded in illuminating for you one of the facets of the wonderful man whom we are gathered here to honor.

Dinner Address

Madame Lecomte du Noüy

When Dr. Shuster asked me to say a few words tonight I thought I could express my thanks to all those who have done so much to organize and contribute to this seminar in memory of my husband. But I was sternly told that there were to be *no* thanks and as during the many years that Dr. Shuster and I have worked together on the Lecomte du Noüy Foundation I have learned to obey him unquestioningly, I will do so now. But he cannot prevent me from *feeling* deeply grateful to you all.

I thought you might be interested in hearing about some of the difficulties encountered by Lecomte du Noüy when he first started to employ physical methods in the study of biological problems. As Dr. Goillot told you this afternoon, my husband had to invent or alter practically all the instruments he used; but I do not think he knows that Lecomte du Noüy made the first model of his tensiometer out of odds and ends found in the Compiegne Hospital, including a piece of string and one of Mme Carrel's hairpins. In those prehistoric times women luckily still used hairpins. It worked well enough to justify his drawing detailed plans and applying for government permits necessary during the war to obtain the materials needed. When they finally arrived he built the first real model himself in a tool shop. Needless to say it was very primitive compared to the final automatic model which has been used the world over. But it worked!

Though a physicist by training, Lecomte du Noüy had become interested in biological problems while working on the cicatrization of wounds and was convinced that physical methods when adapted to biology could bring important results. But biologists in those days had no knowledge of physics and physicists were not interested in biology. As Niels Bohr said when expressing his astonishment that Lecomte du Noüy was devoting his time to it: *Es ist solch eine Schmiere*. Carrel strongly disapproved when my husband told him he wanted to study the blood, saying it was the sewer of the body and he would never find anything. Yet in a few years Lecomte du Noüy had for the first time put in evidence a physical difference between immunized and nonimmunized serum and discovered twelve entirely unknown phenomena.

After the war Lecomte du Noüy worked for seven years under excellent conditions at the Rockefeller Institute. Then the doctors decreed that my health could not stand the New York climate and that I must live in the desert. So we decided to go to Paris! But when Lecomte du Noüy arrived at the Pasteur Institute with all the cases of instruments Dr. Flexner had given him, there was no laboratory big enough to house them, so Roux gave him a big attic. A gabled roof, a rough wooden floor, no partitions, no gas, no electricity!

Lecomte du Noüy made simple plans. Roux agreed to pay the building costs, but everyone was horrified by the innumerable gas and electric outlets that Lecomte du Noüy insisted on having. I helped to unpack the cases, set up the instruments, and gradually learned to use them. From that time on I worked daily with my husband. But many of the old Pasteur scientists were aghast not only at these innovations and by the many unknown instruments but also by my husband's sport clothes, for they still wore the long black capes and black felt hats of Pasteur's time. Even the cleanliness of the department was a subject of criticism. One visitor, known for his untidy laboratory, sneeringly remarked that it was obvious from the spotless condition of the rooms that no work was ever done there; and Lecomte du Noüy's reply that to his knowledge dirt had never added to the accuracy of an experiment did not help to mollify him. At the end of his sabbatical year. Lecomte du Noüy decided to send in his final resignation to the Rockefeller Institute and to stay at Pasteur. The department grew steadily, with students from all over Europe coming to study Lecomte du Noüy's methods and work with him.

When we arrived in the United States from Paris in the beginning of 1943, my husband was dismayed to learn that since we came out of occupied France we were considered suspect and he would not be allowed to work in any war laboratory. The only thing we were both allowed to do was to become air wardens! But one night he was asked to speak about Paris under the Germans, and members of the USO and Salvation Army who heard him asked him to talk on the subject at all the aviation, army, and navy camps. It meant speaking two or three times a day practically every day and traveling in buses at night in between. Lecomte du Noüy had been ill before starting and at the end of a year he was completely exhausted. He was also very anxious to write a synthesis of his three French books which several publishers here had turned down; and as he loved the West we went to stay on a ranch in Montana to write in peace. My husband was a good rider and could not only throw a rope but do trickroping. As cowhands were scarce he helped with the cattle all day and worked on his book evenings and nights. But he had barely finished correcting the proofs in 1946 when he had to be operated on, and he was in the hospital when R. A. Millikan's comments on *Human Destiny* were sent to him. Needless to say they were a big boost to his morale.

Millikan, who had written a preface to Lecomte du Noüy's *Surface Equilibria of Colloïds* published in 1926, had agreed to read and comment on the manuscript but had warned the publishers that though he admired my husband's scientific work this did not mean he would be in accord with his philosophical and religious views. However, his two pages of comments were enthusiastic, and his statement that it was "a book of such fundamental grasp and insight as cannot be expected to appear more than once or twice in a century" launched *Human Destiny*—together, I must add, with Dr. Shuster's favorable review and Fulton Ousler's article in the *Reader's Digest*. *Human Destiny* rapidly made the best seller list, has now attained the highest number of editions of any work of its kind, and has been translated into twenty-three languages. Lecomte du Noüy did not live to know its tremendous success throughout the world. After my husband's death, Millikan—who never passed through New York without coming to see me—repeatedly told me that he always kept *Human Destiny* beside his bed, right next to the Bible, and referred to it constantly.

Some articles written by Lecomte du Noüy between 1929 and 1945 on the evolution of science, the laboratory, aging, evolution, religion, etc., appeared in book form in France in 1964. It was published by David McKay in New York this spring under the title *Between Knowing and Believing*. The American edition has the advantage of being prefaced by Dr. Shuster and of containing two unpublished letters of Teilhard de Chardin written to my husband and myself, in which he expresses his accord with Lecomte du Noüy's point of view. Both editions were illustrated by Bazaine, considered one of France's top modern artists. As his theme for the illustrations of *Between Knowing and Believing*, Bazaine chose the fight between the angel Gabriel and Jacob, thus symbolizing the struggle in Lecomte du Noüy between the scientific agnosticism of his youth and his later spiritual beliefs.

Although written between thirty and forty years ago, these articles are astonishingly up to date even in his concept of how a modern scientific institute should be built. And as Dr. Braisted said to me, were my husband alive today he would still be thirty years ahead of his time. That is why the aim of the Lecomte du Noüy Foundation is to attribute a prize every second year in this country and every other second year in France, to a work which will contribute something to the spiritual thinking of our epoch. The prize so far has gone in a majority of cases to scientists, and this years's prize in France went to the physicist Bernard d'Espagnat for his *Conceptions de la Physique Contemporaine* which is highly thought of by French scientists. Showing as it does the need to introduce the role of consciousness, it tends to break away from a purely materialistic outlook.

For the same reasons, we wanted this seminar to be not only a memorial to Lecomte du Noüy but a discussion and airing of the different subjects he was interested in as they appear today. It is my fervent hope that even if only a small seed is sown it will sprout and that new paths may open which though seemingly divergent will gradually draw closer together until they all finally converge and lead humanity to the ultimate goal of evolution: the truly evolved man, "the perfect man, who is not a myth but who has existed in the person of Jesus."

Today the world looks largely to laws, treaties, social or political theories, and governments for the attainment of universal peace, prosperity, and liberty. But in the final analysis these depend on the men who administer them, and I believe that Lecomte du Noüy was right when he wrote in *Human Destiny* that "peace must be established by transforming man from the interior and not by erecting external structures." A little further on he adds, "To prepare the future by substituting for the individual conscience structures which ignore this conscience constitutes a makeshift doomed to failure and a pathetic waste of time."

As this first day has been consecrated to the life and work of Lecomte du Noüy I hope you will forgive me if I close by citing the last paragraph of *Human Destiny*.

> Let every man remember that the destiny of mankind is incomparable and that it depends greatly on his will to collaborate in the transcendent task. Let him remember that the Law is, and always has been, to struggle and that the fight has lost nothing of its violence by being transposed from the material onto the spiritual plane; let him remember that his own dignity, his nobility as a human being, must emerge from his efforts to liberate himself from his bondage and to obey his deepest aspirations. And let him above all never forget that the divine spark is in him, in him alone, and that he is free to disregard it, to kill it, or to come closer to God by showing his eagerness to work with Him, and for Him.

Reminiscences

Jacques Trefouël

You were good enough, my dear Madame Lecomte du Noüy, to give me the great joy of speaking of the man who was your incomparable companion and one of my best friends. I repeat "the great joy" for I can only evoke him surrounded by the radiant light he cast on everyone as well as on everything he undertook.

A certain evening stands out clearly in my memory. You had been detained for a few days in America by family obligations and we had asked your Pierre, who was always lost without you, to come to the house with a few other friends. As he leaned against the mantelpiece he kept us under the spell of his diversified but always brilliantly intelligent conversation. It was so evocative and teeming with anecdotes that at midnight we were under the impression that we had lived for three hours in a kind of dream from which we tore ourselves with difficulty. We have often recalled that evening which made concrete for me the image of a man blessed with so many gifts: the physique and distinction of an aristocrat, elegant gestures, an unbelievable facility of speech, a prodigious memory, and an apparently limitless culture.

I also see my wife and myself at your property at Clairefontaine, luckily this time with you both. There he was transformed into a kind of cowboy in the bushes, giving me the impression that this was his true setting, only to find him in his laboratory the next day, changed into a serious, concentrated scientist, passionately bent over an instrument entirely the creation of his brain and hands.

None of these aspects of his personality excluded the others. Pierre could be the most urbane man of the world, or a simple being living in love and close communion with nature, or the true scientist whose only object in life was to experiment. That is how I first knew him, in his laboratory at the Pasteur Institute where we spent so many fruitful hours.

You both arrived in 1927, a few years after we did, and what you had accomplished unaided in the rooms on the second floor, bore the mark of your husband's possibilities. Everything was as refined as it was functional. Lecomte du Noüy had adjusted everything himself—an interfacial tensiometer, for instance, a micro-viscosimeter which enabled him to demonstrate that there is an absolute minimum of viscosity for the serum which corresponds to the temperature when certain biological properties are modified. This was the first physical phenomenon correlative to the destruction of the complementary power ever recorded.

He also perfected a turning hydrogen electrode for the determination of the pH—the ionometer, which amongst other things enabled us to study the reducibility of our sulfaminated azo-compounds (antibiotic sulfanilamides).

That month passed in close intimacy further tightened my bonds of affection. Pierre would draw us into your room and our reunions of four remain amongst my most precious memories. He was preparing his book *L'Avenir de l'Esprit*, prelude to the one he was to write later in the United States, *Human Destiny*, which remains the best known of his works, the one which transformed the concepts of thousands of men to whom he brought hope or faith.

Back in Paris, we again gathered together in your home with a few other faithful friends and Pierre maintained our confidence. The courageous activities of both of you were to lead you very far, so far that one day it became imperative for you to leave France. Back in America (after how many vicissitudes!) Pierre had only one desire: to speak of his country and its struggles to as many of your countrymen as possible. He covered thousands of miles, often giving several talks a day, using his strength to the utmost limit. When sickness forced him to give up, it was again with the most total and serene courage that he faced the eventuality of a final departure. "I am the happiest of men and I would not change places with anyone," he said. "I am at peace with God and the world, I have accomplished my task."

How can a man successfully accomplish such stupendous tasks? We must acknowledge that he did so because of the numerous gifts lavished upon him, because of the care he took to develop these, and also because of a circumstance few men can hope to have the benefit of: his meeting with a woman who knew how to help him shoulder his burden and how to share all his aspirations.

Your husband then showed me his shop which exceeded in perfection everything that my ambition or desire could conceive. It was an innovation at the Pasteur Institute which since then has become aware of how deeply he felt the imperious necessity of being able to create personally the tools required for successive experiments. The word "tools" seemed too commonplace, for in this shop belonging to a physicist and an artist the delicate instruments were more like jewels manipulated by Pierre with his expert and shapely hands. His joy in showing them to me equaled mine in admiring them. With his usual kindness he perceived my feelings of envious admiration and suggested that I should use his equipment whenever I wanted. I cannot forget the happiness and gratitude he thus aroused in me and how honored I was by his confidence. I must admit that I often availed myself of his unfailing good will; and the moments we spent together have marked me definitely. I now have my own shop but I humbly recognize that it is only a pale reflection of his, where everything was order, care, and harmony.

The visit to his laboratories included going into a second part, where in a series of rooms behind glass partitions I would perceive your silhouette, swathed and masked in black, replanting the strains of chicken embryos. You were following the avant-grade technique at that time perfected in America by Carrel at the Rockefeller Institute, where the director, Dr. Flexner, had given him and your husband departments with almost unlimited funds at their disposal. The work done there by Pierre marked the beginning of his fame as a scientist, in France especially. It is due to the sound judgment of our beloved Roux that we owed your presence at the Pasteur Institute when your health, Madame, obliged you to leave New York. Roux also liked to follow Pierre's research which can now be considered the fundamental elements which developed, by slow degrees, a branch of molecular biology.

We would occasionally join you during your lunch hour. You brought it already prepared and ate it sitting in the leather armchairs of your office with its sloped attic ceiling—the room which you had made personal and elegant.

In all sincerity I must admit that I was not the only one to seek out Lecomte du Noüy. All those at the Pasteur Institute who were struggling with a difficult problem came to seek its solution from the man who was always happy to speak of science. He led the discussions with youthful enthusiasm, communicating to each one a new ardor. The arguments always ended in restored confidence and good humor.

An astonishing gaiety characterized him. One day he telephoned and asked me to come up to meet Joltrain. We discovered that we had been students together at the Lycée Carnot and reminisced at length. When I took my leave Pierre said seriously, "I thank you for your good visit to my friend Joltrain."

I also remember his admonition, "Beware of people who are eternally unhappy—to me their case seems very suspect." But this apparent lightness, which was his great charm, in reality masked a profoundly meditative spirit.

At the time I knew Pierre, he was devoting his gift of writing to science. In his youth he had moved in literary circles where his outstanding talents had proved singularly varied. He was an author of successful plays, an actor, and an amateur circus performer (as he smilingly reminded me). He lived life fully.

But his exceptional intelligence was to lead him into other fields: oriental languages, political sciences, and law. They all interested him and prepared him to become a philosopher already anxious to buttress his reasoning by scientific arguments. Moreover, his attraction to science was such that at the age of fifteen he had already taken out a license for an invention. Later, he studied under the Curies, Jean Perrin, and Appel. Finally he successfully submitted a thesis at the Sorbonne on the cicatrization of wounds as a function of the age of the subject. These studies in physiology gave him new ammunition.

How can we fail to understand the development of Lecomte du Noüy's research and the harmonious way he juggled with the facts of science and then asked them to unveil deeper secrets? To attempt to catch a glimpse of the meaning of life seemed to him to be his role here below, for himself as much as for others. I followed the progress of his preoccupations with amazement, and in particular during the time when the exodus of 1940 had banished us to the provinces.

Our laboratory had been evacuated to the "Deux-Sèvres," and you joined us there shortly. What would have become of my wife and me without you two, with whom we shared the same despair when listening to the speech of Marshall Petain, and then the same faith in the distant victorious future after another voice had united us four in a common certainty!

The happiness with which you surrounded him, my dear Madame, permitted the full realization of his work, accomplished in enthusiasm and thanks to a community of tastes which is very rare. You lived together in such complete harmony that I do not think I am mistaken in affirming that your country had become a little his own and that your preference went to his.

All those who have approached you both have understood what a union such as yours represented. Lately I reread the lines written to you twenty years ago by Teilhard de Chardin. He reaffirmed the sincere attachment and the complete and unusual communion of faith existing between Pierre and himself and wrote:

> You can surely count that I shall transfer to you all the fidelity and warmth of a friendship of which you have always been a part (I hardly separated you from him in my imagination and my thoughts). But above all may his light in a different but just as real guise illumine your life. Little by little a road will certainly open before you which it behooves you to follow so as to rejoin him in a manner worthy of him, and you can be sure that he will help you. It is action in his memory that will console you.

You have never deviated from that course, my dear Madame, and your desire of love and devotion must be fulfilled by the hours we are living today in memory of him who was one of the apostles of Christ and for you another yourself in intelligence, spirit, and heart.

My Memories of a Great Friend

A. S. Balachowsky

I will try briefly to retrace some personal memories of Pierre Lecomte du Noüy during a rather trying period: the period comprising both the return from the exodus, just evoked by Mr. Trefouël, and that of the Occupation, up to the departure of our friend for the United States. I shall also speak of the period leading to the crowning of his work and to the end of his life in 1947—a life unfortunately much too short for an activity which still held immense possibilities.

Pierre Lecomte du Noüy had a most engaging personality, due as much to his exceptional intelligence and his natural distinction bereft of all affectation, as to his generosity and enthusiasm. He was a "complete man," endowed with a powerful creative intelligence and at the same time with great physical resiliency. He could write all night or spend hours scrupulously controlling a difficult experiment. He could also travel days and even weeks on horseback as he did in the West of the United States, which always attracted him irresistibly and was his country of predilection for the refreshment of his body and spirit.

He was always careful of his appearance, even when employed in manual labor such as farm work. I will speak later of the time he spent on a ranch in Montana during part of the war. He was meticulous about his affairs, his instruments, and even about his clothes, refusing to ignore a spot or the slightest trace of dust. To nobody did he divulge his technique for shining his shoes with a bootpolish of his own invention, the secret of which he never revealed. He also had a recipe for maintaining an impeccable pleat in his trousers, and showed me with pride a suit which was fifteen years old and looked as if it had just come from his tailor.

He was at the same time generous and economical, contenting himself with a frugal meal, but spending without hesitation a large sum of money to buy a new instrument for his experiments or research.

He took great pains to make his style limpid, clear, and easy to read so that the noninitiated could understand what he said about complex physical or biological phenomena. He liked to write with his text illuminated by a single ray of light, in a dark, quiet room, soundproofed and air-conditioned at 20–22°C., which he considered the temperature most favorable for the concentration of thought.

I will not discuss with you Pierre Lecomte du Noüy during the prewar period just described by Mr. Trefouël, nor his career at the Rockefeller Institute and at the Pasteur Institute. Nor do I wish to speak of his fundamental work, for this will be evoked throughout this colloquy. I will simply content myself by complementing the text of Director Trefouël with the fact that I was also part of, and frequented with assiduity, the small circle of intimates that gravitated around the Philomatic Society. Pierre Lecomte du Noüy, who had completely reorganized this society, was its principal animator and always accessible to the young—the young to whom I still belonged in 1932 when I first knew him.

The memories I would like to evoke are more recent and personal and concern the life of Pierre Lecomte du Noüy in the depressing period of the war and the Occupation up to the crowning of his achievement in 1946–1947 in the United States.

After the exodus, in 1940, Pierre Lecomte du Noüy and Madame du Noüy returned to Paris to find, like many other French people, their house of Clairefontaine, situated in a marvelous setting in the Rambouillet Forest, soiled and pillaged by the invaders. They abandoned and would never again inhabit the "Pavillon," as their house was called. A community of sentiment and thought during this somber part of our history drew us closer together in a few months than would have years of friendly scientific relations.

Lecomte du Noüy had organized meetings on Sunday afternoons in his apartment on the Rue de Grenelle, from where the Golden Dome of the Invalides could be seen silhouetted against the sky. Scientists, writers, philosophers, and doctors gathered there and continued an intellectual life in the brutal obscurantism of the Hitler occupation.

The Occupation had become unbearable to us both and in the paralysis of the last months of the year 1940, our hatred of the occupying power, shared by Lecomte du Noüy and myself with the same force, had become a powerful uniting link. Little by little the circle of friends and intimates was weeded out, as the two contrary currents soon to divide the French became more pronounced every day. There were those who could already be called "those of the Resistance," and those who, after Montoire, were to become "the collaborationists." The break between Pierre Lecomte du Noüy and Alexis Carrel, in spite of thirty years of collaboration, dates from that period, for Carrel had embraced the cause of "the New Europe."

We both listened to the BBC broadcasts in English for they were never jammed and we commented on them together. All the broadcasts of General de Gaulle were heard, and on November 11, 1940, we were on the Champs-Élysées when the Vichy police arrested the first patriots who displayed red, white, and blue cockades.

In 1941, the weight of the Occupation became daily heavier, more menacing, more cruel. At the same time the names of the first victims of the internal Resistance appeared more and more frequently on the yellow posters of the "Kommandatur" and among them that of our colleague Holweck of the Pasteur Institute. He was tortured to death, and his bleeding corpse was sent to his family "as an example."

Meanwhile, I had even at that early period established the first contacts with London, and we had been asked for information about the movement of troops and the fortifications of the coast. Pierre Lecomte du Noüy was able to give such information with great precision about certain villages on the Norman coast which were luckily beginning to be cleared of the "shock troops" after the defeat of "Operation *Seelöwe*" (the invasion of England).

In September 1941, I suggested to Pierre Lecomte du Noüy that we should "escape" for about ten days and go down the Loire in a kayak. He was enthusiastic about the idea and we left by train from Orleans with eighty-eight pounds of equipment. The kayak was quickly assembled on the borders of the Loire, the sleeping bags, tent and provisions were stacked in the bow and in the stern and we left quietly, paddling with the current towards Beaugency, Blois, Amboise, and Tours, through the most beautiful country of France, witness of many other invasions. We needed a circulation pass, we did not have any, and camping out of doors was *streng verboten*, but I knew some remote isles in the stream, covered by vegetation, difficult of access from the banks, where we could establish our camp in peace.

During this long week of a "clandestine holiday," favored by a beautiful summer's end, everything seemed to live in peace in the midst of a tranquil nature teeming with birds, small animals, and especially rabbits which were no longer hunted and had not yet been contaminated by the myxomatosis of Dr. Armand Delille. During eight days we met practically no human being, either military or civilian. Only a few disintegrated carcasses of abandoned cars on the shores reminded us at intervals, together with the broken down bridges, of the battles which had taken place a year before in this landscape of "la Doulce France" so little made for war.

It was during our vigils under the tent that Pierre Lecomte du Noüy for the first time evoked the project of a synthesis which was to take shape after his trilogy: *L'Homme devant la Science* [The road to reason], *Le Temps et la Vie* [Biological time], and *L'Avenir de l'Esprit*.

His scientific formation, being more that of a physicist than of a biologist, had led him away from the natural sciences towards which he was now irresistibly attracted and was sedulously studying. The problems of evolution, of the evolution of man in particular, totally occupied his mind.

We arose and retired with the day. In this calm nature, where only the current of the millenary river broke the silence of the night with its regular murmur, we exchanged our thoughts and our hopes, while snuggling in our sleeping bags.

This intimacy with an elite being who had such a wide experience of life, of science, of men and things, and who had roamed over the entire world, aroused an amazement in me which was most beneficial. We navigated from isle to isle, from encampment to encampment, and I acquainted him with the birds, insects, and plants. Everything interested him.

"The other day," he wrote me in 1946, "we assisted at the birth of a red dragonfly emerging out of a horrible black grub. The slow development of the body, the wings, the pigmentation, slow but astonishingly rapid (about a dozen hours), stupefied us."

This life of physical relaxation coupled with frugal nourishment did us both the greatest good and stimulated us physically and intellectually. But we had to go back to Paris, and thereafter we often evoked this time spent together which had further sealed our friendship, if there had been need.

On December 7, 1941, after the Wehrmacht had got bogged down on the plains of Russia, came the blow of Pearl Harbor. Germany was destined to be crushed in a vise by the two greatest powers of the world, but we were still far from the final victories.

Lecomte du Noüy felt himself condemned to inaction. A torrent of generous, useful, and creative ideas was welling up within him that he could not put into execution. He decided to leave for the United States, but the distance between *Gross Paris* and the Statue of Liberty was long. I strongly advised him to leave Europe so as to make known in America all he had already lived through and seen of the Hitler occupation and to tell the Americans that thousands of Frenchmen lived in the expectation of taking up the fight at their side with the means of despair remaining to them.

The year 1942 arrived with its bereavements, its devastations, its massive round-ups, its executions, its deportations. Yellow stars appeared on the dresses and garments of women and children belonging to a race declared accursed and doomed to disappear. I remember when Dr. Etienne Bernard, during one of the Sunday meetings, came into Lecomte du Noüy's living room wearing his new yellow star. With tears in his eyes, Pierre seized him with both arms and embraced him in front of the deeply moved onlookers.

Pierre and Mary Lecomte du Noüy passed into the "Free Zone" in July 1942 and obtained their visas for Portugal thanks to the help of Jacques Trefouël and Dr. Aublant, and to a little extent myself through members of the NAP *reseau* (Resistance Group Infiltrating Public Administrations). After a thousand difficulties they left Europe on December 16, 1942, on an overladen, uncomfortable boat which took a month to reach Philadelphia. A new life was to begin for them but alas Pierre Lecomte du Noüy was never to see Europe again. He had, moreover, felt a foreboding as the shores of Portugal faded on the horizon before his eyes.

America at war was not comparable to occupied France. Food was abundant, civilian life continued almost normally, but this outward calm hid an intense activity with the nation's nerves tensed twenty-four hours a day in the greatest military effort in history.

From the time of his arrival, Lecomte du Noüy served the cause of France tirelessly, and under the sponsorship of the YMCA he delivered hundreds of talks to young recruits in the military training camps. After June 6, 1944, many of these young men were to sacrifice their lives to liberate our country from the most horrible of all dictatorships.

Lecomte du Noüy covered the whole country from East to West at the price of great fatigue, traveling ceaselessly, giving up to four and five lectures a day. His efforts were repaid by his success with the "guys" to whom he spoke of what he had seen in occupied France as well as of French culture and philosophy.

Lecomte du Noüy also undertook to work on a ranch in Montana, for there was a great shortage of hands. He led the life of a cowboy, in the saddle from morning till night. He knew how to brand cattle, herd them into the corrals, and how to throw a rope with the dexterity of a westerner. But at the end of these exhausting days instead of letting his body rest he gave himself up to intellectual work and began to write the first pages of his new book, *Human Destiny*.

The year 1944 came to an end, and with 1945 the Allied victory became definitive. I landed in New York on the *Clipper* on June 11, 1945, after two years spent in Buchenwald. We had been liberated only a month and a half before by a group of Patton's army to whose sacrifice my comrades and I owed our lives. The contrast was startling. André Malraux, the minister of information, asked me to leave for the United States and give a series of lectures in the universities. He had two precise aims in view. First I was to explain the French Resistance and its organization to the Americans and thus prove to them that we had never given up the fight. Secondly, I was to make known in America the horrors of the Nazi regime and specifically the experiments on human beings used as guinea pigs at which I had assisted in Buchenwald and about which I was to testify at the beginning of 1946 during the trial of the great war criminals in Nuremberg.

On hearing of my arrival in the United States, Pierre and Mary Lecomte du Noüy invited me to come and stay two weeks with them at the Tee-la-Wucket ranch in southern Colorado. It was in this marvelous setting of the Rocky Mountains in the Hopi Indian country with its grandiose Walt Disney type scenery that I was able to free my body and mind from all the intoxication accumulated in the course of the dramatic years passed in a disrupted and ravaged Europe. Lecomte du Noüy was finishing the manuscript of *Human Destiny* and during our horseback rides or in front of a succulent barbecue, where a whole quarter of beef was roasting, we untiringly discussed its theme. After two weeks of this real holiday I retook the road for California through the New Mexico desert and stopped at the Grand Canyon to admire one of the most beautiful sites in the world.

I saw Lecomte du Noüy again in Los Angeles and in New York before leaving America in 1946. I was never again to see my great friend. An exchange of letters perpetuated our friendship and I followed the immense success of his book.

The first news I received was optimistic. Lecomte du Noüy had settled in Altadena, California, ten miles from Pasadena, at the foot of Mt. Wilson, in fairylike surroundings and an ideal climate. There he concentrated all his efforts and devoted his time to the works of philosophical science to which he had dedicated the remainder of his life.

On May 15, 1946, he wrote to tell me how much he regretted my absence and asked me to come and join him. "Try to obtain a mission for this summer. In the garden there are palms of every variety, avocado trees, guava and pomegranate trees, peach, cherry, apricot, apple trees, etc . . . an astounding orchid tree 15 meters high, cacti, hibiscus, goodness knows what else!"

He continued:

> I await the proofs of *Human Destiny*. The publishers were so enthusiastic and the notice they prepared was so startling I was obliged to recommend prudence. I am very curious to see the reaction of the public. We have been so busy getting settled that we have not yet seen anyone. In this setting one can dispense with the rest of the world. Ah, if you could come. . . . Tell me about the organization of your laboratory and your plans. You had spoken of an African mission. What has become of it? We would prefer to have you come here! And now I must do some watering or cut some branches in this three acre jungle. I get up at seven o'clock every morning—an unheard of thing for me—and I never have time to do everything.

The disease which was to carry him off was unfortunately gaining steadily and manifested itself brutally at the end of 1946, leaving little hope. Lecomte du Noüy fought with all his tenacious energy to give to life and men his last messages of optimism. He still gave talks and commentaries on the radio. *Human Destiny* had become a best seller, and was to be translated into all languages. But at the very time when he was in request everywhere, when triumph knocked at the door, his strength left him. In spite of excruciating pain, he had himself transported on a stretcher to a radio station so he could speak to men–to mankind, whose betterment he longed for.

On April 27, I received this last letter:

> Yes, if success could satisfy me I would be happy, but I have reached the age when it no longer counts. What counts is the survival of constructive ideas, the ascending march to true progress, Evolution and who knows what will remain of it.
>
> Here, where everything is on another scale, the success of *Human Destiny* is great; 60,000 copies sold in two months. "Best Seller" in the United States, and alas, fifteen letters to answer daily. Now they expect to pass the 100,000 mark. I can hardly stand up. I weigh less than 103 lbs. I look as if I had come out of a concentration camp. . . .

Lecomte du Noüy remained riveted to his bed of pain.

I would like to quote a last message that he wrote only a few days before his death to the world for a book by Thomas Hewes. It was quoted by Madame du Noüy in the admirable book she wrote about her husband:

> Freedom, like health, is only understood or valued by people who are deprived of it and then it is usually too late. The trouble is that

> the death of liberty is often an imperceptible but progressive process like the slow rise of a river, and the people who are only concerned with their own individual problems are too busy to understand that eventually they are bound to be personally involved. Any form of government which does not respect human dignity, obviously based on liberty, is sure to drift into some form of totalitarianism and to crush individual initiative.

No prophecy was ever truer in the period we were then living; but it is equally true for all times.

On September 22, 1947 at 7:30 in the morning this great intellectual and noble figure breathed his last, holding the hand of Mary Lecomte du Noüy who had not left him during the course of his painful illness.

Commemorative Greetings

Paul J. Braisted

It is indeed a privilege which you, Madame du Noüy, have provided us on this occasion. We have come honoring the memory of Pierre Lecomte du Noüy. We owe him much and welcome this occasion to reflect upon his life and his varied activities. In anticipation of these days I have been rereading, as perhaps you have also, his books and essays. As a result I found much to admire and ponder. But beyond all that, I have one paramount impression. It arises as a fresh and vivid realization of the relevance of his life and message. It is on this that I shall make a few comments.

We shall be seeking in various ways to know the meaning of this life of reason, of imagination and of faith. We shall see more clearly what it meant in his own time. We shall also do well to reflect upon the significance of that life and work for us in our time, for men of other lands and traditions and for men of tomorrow. For in a larger and in a truer sense the significance and relevance of his life increases with the passing years.

Pierre Lecomte du Noüy is relevant for our time first of all because of the fullness of his life and thought. His scholarly life was not confined merely to some specific inquiry, nor was his writing and teaching restricted to a narrow field of vision like so much of the academic discourse in our time. His inquiring mind reached out not merely to understand a particular natural phenomenon, but also to understand nature and man's relation to it. For him, an advance in knowledge was not merely an achievement, a new and better understanding, but it was also a new frontier, a call for renewed high adventure of the mind. Devoted as he was to the rigorous work of the laboratory, he could not have been confined to the narrow specializations which too often characterize academic life. His life and thought transcended the barriers which separate or are alleged to separate those who work in the field of science and those who work in the field of humanities, and the superficialities and trivialities of recent debates about the "two cultures" would surely have seemed trivial to him. For him, the life of reason and faith were not two worlds in conflict, but aspects of the whole human experience. In a very special sense he was a complete man, and so a challenge to us all.

Pierre Lecomte du Noüy is also relevant to men of many lands and many traditions in these days. This is not only because *Human Destiny* has been translated into other tongues and has been widely distributed abroad. It is not only because it is today one of the popular books in student book stores in all our universities, read by our own students and by foreign students as well. Rather it is because he recognized the essential unity of men of faith, of all traditions and all times. In his essay "The Future of Spirit," written in 1941, he spoke of "fundamental moral rules–universally shared by all civilizations which have existed in writing for the past six thousand years and have been followed for a much longer time." This is the wisdom of a man who lived and wrote before the recent rebirth of the ecumenical spirit within and among the churches. He was a scientist, and science itself knows no national boundaries. A new atmosphere of understanding is growing up between leaders of differing religious and cultural traditions in various parts of the world. Undoubtedly there are some here today who have experienced a sudden awakening to a sense of unity with men of another culture or another faith. In such moments we have become aware of a unity in a deeper life and one that we recognize in spite of differences in tradition or language or systems of thought by which men have sought to communicate the sense of a deeper and fuller life. These are surely token of a future yet to be, but perhaps not so distant as is often thought. It is foreshadowed by the words of an American philosopher lately honored by the Du Noüy Foundation, William Ernest Hocking, in his volume entitled *The Coming World Civilization.* He concluded his discussion of the future of faith and knowledge by saying that

> with the relations between religions relieved of confusion at once by the growing unity of their unlosable essences, the understanding acceptance of variety and the quiet convergence of purpose in the identity of a historic task, religion will be able to bring a new vitality into the disturbed motivations of mankind.[1]

These words of Dr. Hocking seem to me to continue a quest foreshadowed in various affirmations of Lecomte du Noüy, and the future development that he would have understood and encouraged.

And finally, Pierre Lecomte du Noüy is also relevant for our time and tomorrow because what he called "the moral crisis which has prevailed since the beginning of the century" continues in aggravated forms. The turmoil and tragedy which he witnessed, are in some ways only a pale reflection of the fears that drive mankind now, of the violence and slaughter which characterize so much of contemporary life. As he spoke out of faith with a clear conviction about the future even in the shadows of his day, does he not also speak in the seemingly deepening shadows of our time? And when he speaks of spirit, it is not to offer a substitute for reason and knowledge, but as a part of and fulfillment of man's destiny in a larger unity of life. In a memorable essay to which I referred above, written twenty-six years ago, we find these lines:

> I have confidence in the future of spirit, in the grandeur and nobility of the part that man is free to play. The struggle for existence and for evolution of which he is the crowning point continues, and the fight has lost nothing of its violence by being transferred from the material round to the spiritual round. The dignity of the individual must emerge from the effort he makes to break loose from the yoke of the flesh and obey the inner voices.[2]

Such faith is found not in avoidance of the tumult and sickness of contemporary life, but is rediscovered again and again by every pioneering spirit who is not content with things as they are, but presses on toward the future that must be. This task has been expressed by another scientist-scholar recently honored by the Du Noüy Foundation, Loren Eiseley, in his volume, *The Firmament of Time*. He wrote:

> Let it be admitted that the world's problems are many and wearing, and that the whirlpool runs fast. If we are to build a stable cultural structure above that which threatens to engulf us by changing our lives more rapidly than we can adjust our habits, it will only be by flinging over the torrent a structure as taut and flexible as a spider's web, a human society deeply self-conscious and undeceived by the waters that race beneath it, a society more literate, more appreciative of human worth than any society that has previously existed. That is the sole prescription not for survival, which is meaningless, but for a society worthy to survive. It should be in the end a society more interested in the cultivation of noble minds than in change.[3]

It is in this broad perspective of yesterday, today, and tomorrow that we can begin to see and know the continuing and growing relevance of the life and message of Pierre Lecomte du Noüy. This is the challenge of one whose faith and work we commemorate in these days. His spirit still beckons men of science, of imagination, and of faith.

NOTES

1. William Ernest Hocking, *The Coming World Civilization* (New York: Harper and Brothers, 1956), p. 185.
2. Pierre Lecomte du Noüy, *Between Knowing and Believing* (New York: David McKay, 1966), p. 237.
3. Loren Eiseley, *The Firmament of Time* (New York: Atheneum, 1960), p. 7.

The Scientist's Mission

Pierre Leroy, S. J.

The mission which weighs upon the scientist is a serious and grave one. If it is his personal duty to analyze the natural or experimental mechanisms which offer themselves for his investigation and in this way to enrich humanity, it is also incumbent upon him to express their deeper meaning.

In one of his philosophical essays, Einstein makes a point of this special vocation of the scientist. "Individual existence," he writes, "is for him a prison and he wishes to live in possession of the fullness of all that exists in all its unity and its ultimate meaning. . . . Serious scientists are the only men who are seriously religious."[1]

The demands which the scientific method makes create, no doubt, limits which one cannot cross without engaging oneself in a field which is not one's own. If the scientist allows himself to make a value judgment he runs the risk of criticism and of losing the prestige he has among his colleagues because of his scientific knowledge: Science based on rationalism does not allow the scientist to go outside its limits. Here I am speaking of totalitarian rationalism which is contrary to the spirit of science and which according to modern physicists no longer has any place nowadays. True rationalism does not a priori set limits to that which may exist. As was recently pointed out by Bernard d'Espagnat this is a *method*, not a *philosophy*. It is therefore wise to think before condemning attitudes of mind which seem at first to conform but little to rationalist demands.

Who among us has not regretted, however, that so many men who are famous for their scientific knowledge and competence have remained silent when faced with the major problems of human destiny? And how happy we have been, on the other hand, when we have heard the wise and prudent reflections of some of these superior minds. The desire to transmit their vision of the world to others is the principal preoccupation of those men who feel that the star which illuminated their existence is setting and will have no peace until they have offered the fruits of their meditations.

Pierre Lecomte du Noüy was one of those who fulfilled this twofold mission: a scholar in mathematics, physics, and biology, fascinated to the last by the phenomena of living things, devoted more than any other man of his time to the study and consequences of the organic evolution. He did not hesitate towards the end of his life to launch out into the least accessible fields.

Conscious of the basic insufficiency of descriptive science to understand the nature of reality, he realized that a philosophy of nature and of its culmination *man* could alone complete a picture which, without him, would be incomprehensible.

Reality is all-embracing, containing at the same time that which is organic and determined and that which is spontaneous and free. In order to grasp all that this implies, Pierre Lecomte du Noüy proposes a hypothesis which in the estimation of certain of his colleagues would have been sufficient to disqualify him had they not been impressed by his strong personality and his conviction.

The fundamental facts admitted by all biologists are grouped together and resumed by him in the following manner: life begins to manifest itself in simple beings; gradually it evolves into more and more complex forms until it finally reaches the human brain. From this, moral and spiritual ideas spring forth whose spontaneous development spreads throughout the whole world. He states, moreover, that even the most comprehensive and liberal science is unable to explain these facts and their evolution.

It is therefore essential to find a hypothesis which is in keeping with the correlation of these findings and which links them to inorganic evolution so as to give a significance to the whole. Some sought to oppose "chance" with "anti-chance," a word created for the needs of a cause. It would have the merit, if one accepted its existence, of orientating "the series of phenomena in a progressive direction which is highly improbable," as Pierre Lecomte du Noüy wrote in *Human Destiny*.

Pierre Lecomte du Noüy does not wish to build on such an imprecise foundation; he turns to classical philosophical data and to religious faith to bolster the profound reasons for evolution. The hypothesis which he proposes is "a telefinalism." It is an attempt to explain the movement towards a distant goal. It lays down the principle that scientific laws established by man for matter, insofar as they are coherent, correspond to an objective reality. It is based upon the impossibility of attributing the beginning of life, its evolution and the manifestations of mental activity, simply to chance.

I am well aware that this hypothesis has been criticized, for the mathematical demonstration on which it is partly based is not convincing. In any case we must not see this as an attempt to rehabilitate a narrow minded finalism which claimed to be as D'Armagnac suggested the expression of a "particular intention in every object". This interpretation would be a betrayal. In fact it is a matter of finding a significance in that evolutive movement which reaches its climax finally in the human brain. I do not think I am going further than the thought of Pierre Lecomte du Noüy in seeing in the successive arrangements which lead from inorganic matter to man, a polarity, an orientation of a movement towards a distant goal.

True finality only makes sense with man; at his level, choice, intention, and liberty determine action. At this stage of evolution therefore we find the risk of individual liberty which is full of unexpected events, refusals, daring steps, and contradictions but which, at the same time, is the most precious jewel which man possesses.

"Everything has taken place as if Man had been willed," writes Pierre Lecomte du Noüy in *Human Destiny*.

> Man with his present brain does not represent the end of Evolution but only an intermediary stage between the past . . . and the future. . . . We are free to destroy ourselves if we misunderstand the meaning and the purpose of our victories. Our freedom of which we may be justly proud, affords us the proof that we represent the spearhead of Evolution; but it is up to us to demonstrate by the way in which we use it, whether we are ready yet to assume the tremendous responsibility which has befallen us almost suddenly.

If we can judge from what is happening around us, it would seem that few of our contemporaries have followed this example. Let us not be deceived. The apprenticeship of our liberty tends toward a sublimation in true liberty which so many good men, believers or otherwise, are looking for obscurely. We shall never guess by what secret ways the "Dieu Createur" of Pierre Lecomte du Noüy teaches us to purify and to liberate ourselves. "We remain a mystery to ourselves until death," writes Julian Green. "We face ourselves like strangers who look at each other without understanding." It is nevertheless true that this work is accomplished in the obscure channel of evolution and in the progressive discovery of unlimited possibilities. "To be sure, we cannot be too rigorously severe at our stage of Evolution," Pierre Lecomte du Noüy writes in another place. "We are at the beginning of the transformations which will require a sustained effort for hundreds of centuries."

Madame, may I be permitted at the end of this first day of our symposium, to express our sincere and respectful thanks. The speakers who have addressed us so knowledgeably on the works of Pierre Lecomte du Noüy have set vibrating, thanks to you, the harmonics of a fundamental note which is struck by the work of the great and famous man of science who was your husband.

The benefit which we owe to you will, no doubt, allow us—and this is my sincere wish—to live more intensively the ideal which was bequeathed to us and which is the glory of man in the face of his destiny.

NOTES

1. Albert Einstein, *The World As I See It* (New York: Philosophical Library, 1949), pp. 26 and 28.

Pierre Lecomte du Noüy and the Conscience of Our Times

Henri Mavit

Pierre Lecomte du Noüy is an integral part of our culture and of the conscience of our times not only through his works, but also through his personality. His books have been translated into more than twenty languages. In *Rayonnement de Lecomte du Noüy*, Jean Huguet writes that in his village of Vendée, an old, self-taught peasant had shown him a copy of *Human Destiny* among the works that inspired his spiritual development. As president of the "Literary Youth of France," Jean Huguet is well qualified to know what Pierre Lecomte du Noüy can bring to the younger generation. His influence transcends intellectual circles, bringing comfort to those who without it would have lost all hope, as in prisons where men have rediscovered a meaning to existence by reading it.

The fact that his influence has made itself felt in a great variety of circles over the whole world is probably due to the exceptional breadth of his personality. Rarely have so many gifts been united in a human being. An all-round sportsman, expert horseman and roper, he could compete with the best riders of the Far West with whom he had fraternized in the course of his youthful journeys. A gifted actor both on the stage and in the movies who had refused tempting contracts, he was also a writer who could have dedicated himself to a literary career. Guy de Maupassant and Sully Prud'homme were friends of the family and encouraged his first steps. In 1910, his play *Maud* was given at the Odéon and we still admire it. Avid for knowledge, he followed the courses of the School of Oriental Languages, passed his law examinations, and was also passionately interested in philosophy. His love of physics and extraordinary mechanical ingenuity had already revealed themselves at an early age.

In the end Pierre Lecomte du Noüy consecrated himself to science and worked for thirty years in the laboratories of the Rockefeller and Pasteur Institutes. The scientists associated with him have borne witness to his scientific ability, combining inspiration with methodical rigor and a prodigious manual dexterity. We were all astounded by the instruments he invented and the drawings he made with such precision when they were shown and so ably explained to us on the opening day of this colloquy. When I was a student at the Faculty of Medicine and Pharmacy of Marseilles in 1934, our professor of physics, Rimattéi, had expressed his interest and admiration for the experiments of Lecomte du Noüy on differential tensions. Some of his work on the rate of cicatrization of wounds has a philosophic significance and aroused the interest of Bergson who wrote him:

> I have just read *Biological Time* which you were good enough to send me and I wish to tell you how much your book has interested and taught me. Your experiments and your general views on cicatrization which form its main thesis would alone suffice to make your book an important one. But you did not stop there and on these precise observations and reflections you have grafted a new concept of physiological time which I believe is sound and fruitful and which leads you to interesting considerations on time in general. All my congratulations.[1]

His experience as a scientist was to lead Lecomte du Noüy to reflect on science and its methods and to undertake a deep examination of intellectual conscience. He discarded a closed science in favor of an open science. The pretentions of scientism and mechanism seemed to him unfounded. The abuses of the reducing analytical method lead us to a dissociated universe without significance, deserted by life. Moreover, does not the very progress of science broaden the points of view and instigate a rejection of negative dogmatism?

From the study of physics he proceeded to a vaster quest relating to evolution as a whole. If there is evolution, there must be a tendency to evolve. This tendency cannot be explained by chance alone. Moreover the impetus seems to have a direction. It rises towards more conscience to result in man. The latter no longer evolves physically but liberty appears with him as does the esthetic sentiment and the realization of the "beyond." He becomes responsible for an evolution which continues through the spirit.

The Road to Reason was written on the eve of the Second World War, *L'Avenir de l'Esprit* during the Occupation, and *La Dignité Humaine* just after the Liberation. The works of Lecomte du Noüy are an answer to the anxiety of the contemporary world and a commentary on the drama of our epoch in which the errors of intelligence and its deviations have their share. The criterium of morality and of action consists in respect for the human person. "The State should be the servant of Man. Any government which seeks to substitute its interests for the pursuit of individual development is regressive and threatens human dignity." Though he is optimistic about the direction of evolution, he remains apprehensive for the near future. Humanity is torn by the conflicts engendered by frontiers. Lecomte du Noüy is deeply aware of the perils it runs and anxious to make them known. We know the part he played in the Resistance. He was both a man of action and a thinker. His aim in writing his books is to contribute to the advent of a world that is truly free. Science alone will not suffice to bring this about. Even if it is related to spirituality by its inspiration, there is more in man.

Meditation on science and evolution had thus led him to propound the hypothesis and the problem of God. He could not content himself with "anti-chance." In his eyes the mission of the scientist should be to reconcile the rational and the spiritual. This search for unity is also an essential aspect of our time whose tensions and aspirations he shared intensely. He began by fighting a stifling and sterile scientism that he saw at work in the circles he frequented as a result of his scientific vocation. A dogmatic narrowness and an incomprehension of anything religious also had to be overcome. He was one of those who foresaw a spiritual renewal and the unlooked-for points of similarity in ideas which are its condition. Necessities of reason as well as of the heart! Lecomte du Noüy was for a long time an agnostic. Renan was then perhaps the author he quoted most frequently, the Renan of the preface of *L'Avenir de la Science*.

One of the characteristics of this thinker, one of the signs of his generosity, is the satisfaction he feels in discovering relationships and affinities. In this respect we must first of all mention the physicist Charles Eugene Guye, whose principal work "Physico-Chemical Evolution" he often quotes. *L'Avenir de l'Esprit*, by Lecomte du Noüy, *Invention et Finalité en Biologie* by Lucien Cuénot, *Le Message Social du Savant*, by Rémi Collin, *L'Oeuf et son Dynamisme Organisateur* by Dalcq, were all published in the same year: 1941. In spite of their differences these works all showed a similar spirit of method and preoccupations. The spiritual itinerary of Charles Nicolle, who passed from pure rationalism to a larger spirituality and to faith, is analogous to that of our author. After writing *La Morale et la Nature en Biologie*, Charles Nicolle had divulged his research and personal meditations in his book *La Destinée Humaine*.

Finally, how can the telefinalism of Lecomte du Noüy, his vision of man at the summit of evolution, the direction of the latter fail to evoke the name of Teilhard de Chardin? Though the latter read Lecomte du Noüy's books when they were published, the two men never discussed their ideas, and met once for a few minutes on the boat which was bringing them back from the United States. Their incontestable intellectual and spiritual kinship does not prevent occasional differences of points of view and interpretation.

On the one hand, a religious scholar and mystic; on the other hand, an agnostic and a man of science who only rediscovers God after a long journey! Lecomte du Noüy points out very precisely all the differences that separate inert matter from organized matter. With Teilhard, though there are thresholds, the movement of complexification enables one to pass from matter to life. Lecomte du Noüy is more mistrustful of socialization if we take into consideration only the present and the relatively near future. There is in him a real anxiety concerning the supremacy of the social over the individual. But this is a very schematic view if we take into consideration the scope of Teilhard de Chardin's work.

When Lecomte du Noüy's ideas are compared to those of several contemporaries, one is struck by certain utterances of Jaspers, who was himself a scientist before being a philosopher: "The scientific spirit will shine with a purer lustre in proportion to the degree that the scientist becomes conscious of the limits of knowledge and will know how to leave the field clear for other sources of truth in the hearts of men."

Lecomte du Noüy considers the role of the individual of prime importance, and Jaspers says:

> When the individual is not conscious of the fact that things depend precisely on him, when he does not act as if the principle of his actions should be the very principle of the world which remains to be created, it is the liberty of all which is lost. That is why it is the duty of each individual not to abandon himself to the dogmatism of sociological, psychological or racial fatalism nor to the chaotic disorder of life.

I will add this other thought of Jaspers, with which Gabriel Marcel will certainly agree: "Man does not let himself be reduced to mere knowledge."

The great physicist Max Born, asking himself how humanity can render itself capable of utilizing the acquisitions of science without committing suicide, declared: "It should suffice to take the teachings of Christ seriously and to judge good and evil no longer from a national criterion but from a human criterion."

It is one of the signs of our time that we ask for a work to be justified not only by reasoning but by our existence. Perhaps that is why we interrogate scientists even on subjects which go beyond their specialty. We tell ourselves that their experience and their effort of research gives more value to their thoughts. Lecomte du Noüy was conscious of this responsibility of the scientist. Abandoning the silence of his laboratory, he undertook to communicate the results of his meditations. This was the task of his last crowded years in the United States where in numerous lectures he sowed his message for the future. When exhaustion set in, an internal force compensated for that of the body. In the face of suffering, at the approach of the supreme test, his spirit radiated ever more intensely. It is then that he truly saw the merciful countenance of Christ.

We can apply to his example the lines he wrote. They did not refer to him but have the value of a testimonial:

> We cannot but be struck by the disproportion between the duration of a man's life and the duration of his influence on future generations. The destiny of Man is not limited to his existence on earth and he must never forget that fact. He exists less by the actions performed during his life than by the wake he leaves behind him like a shooting star. He may be unaware of it himself. He may think that his death is the end of his reality in this world. It may be the beginning of a greater and more significant reality.[2]

NOTES

1. Mary Lecomte du Noüy, *The Road to "Human Destiny"* (New York: Longmans, Green and Co., 1955), pp. 187–188.
2. *Ibid.*, p. 269.

BIBLIOGRAPHY

List of Books by Pierre Lecomte du Noüy

1926. *Surface equilibria of biological and organic colloids.* New York: Reinhold.
1929. *Equilibres superficiels des solutions colloidales.* Paris: Masson.
1933. *Méthodes physiques en biologie et en médecine.* Paris: Lib. J. B. Baillière.
1936. *Le temps et la vie.* Paris: Gallimard.
1936. *Biological time.* London: Methuen. 1937. New York: Macmillan.
1936. *La température critique du serum.* Paris: Hermann.
1939. *L'homme devant la science.* Paris: Flammarion.
1942. *L'avenir de l'esprit.* Paris: Gallimard.
1944. *La dignité humaine.* New York: Brentano. 1952 Paris: La Colombe. 1967. Paris: Fayard.
1945. *Studies in biophysics: the critical temperature of serum.* New York: Reinhold.
1947. *Human destiny.* New York: Longmans, Green & Co. (French edition, 1948. *L'homme et sa destinée.* Paris: La Colombe. 1967. Paris: Fayard.) Translated into twelve other languages.
1948. *The road to reason.* New York: Longmans, Green & Co.
1964. *Entre savoir et croire.* Paris: Hermann.
1967. *Between knowing and believing.* New York: David McKay.

List of Scientific Works

by Pierre Lecomte du Noüy

1. Cicatrization of wounds. Mathematical expression of the curve representing cicatrization. *J. Exp. Med.* Vol. 24, pp. 451–460, November 1, 1916.
2. The relation between the age of the patient, the area of the wound and the index of cicatrization. *J. Exp. Med.* Vol. 24, No. 5, pp. 461–470, November 1, 1916.
3. Du rôle relatif de la surface et du périmètre dans le phénomène de la cicatrisation des plaies et de la formule qui les interprète. *C. R. Ac. Sc.* t. 164, p. 63, January 2, 1917.
4. Mathematical study of the extrapolation formula and the curve of cicatrization. *J. Exp. Med.* Vol. 25, pp. 721–728, May 1917.
5. Influence on the healing of wounds of variations in the osmotic tension of the dressing. *J. Exp. Med.* Vol. 26, No. 2, pp. 279–295, August 1, 1917.
6. Recherches expér. et applic. des méthodes de mesure et de calcul à un phénomène biologique: la cicatrisation (Gauthier-Villars). *These Ac. Sc.* Paris, December 18, 1917.
7. Recherche d'une équation générale de la loi de cicatrisation normale des plaies en surface. *C. R. Ac. Sc.* t. 167, p. 39, July 1, 1918.
8. A general equation for the law of cicatrization of surface wounds. *J. Exp. Med.* Vol. 29, pp. 329–350, April 1, 1919.
9. A new apparatus for measuring surface tension. *J. Gen. Physiol.* Vol. 1, No. 5, pp. 521–524, May 1919.
10. Energy and vision. *J. Gen. Physiol.* Vol. 3, No. 6, pp. 743–764, July 20, 1921.
11. Latent period in the cicatrization of wounds. *J. Exp. Med.* Vol. 34, No. 4, pp. 339–348, October 1, 1921.
12. Sur l'équilibre superficiel du sérum et de quelques solutions colloidales. *C. R. Ac. Sc.* t. 174, p. 962, April 1922.

13. Spontaneous decrease of the surface tension of serum. *J. Exp. Med.* Vol. 35, pp. 575–597, April 1, 1922.
14. Action of time on the surface tension of serum solutions. *J. Exp. Med.* Vol. 35, pp. 707–735, May 1, 1922.
15. Sur l'équilibre superficiel du sérum et de certaines solutions colloidales. *C. R. Ac. Sc.* t. 174, p. 1258, May 1922.
16. Recovery after lowering by surface active substances. *J. Exp. Med.* Vol. 36, pp. 115–134, July 1, 1922.
17. Action of temperature (surface tension of serum). *J. Exp. Med.* Vol. 36, pp. 547–558, November 1, 1922.
18. A new viscosimeter. *J. Gen. Physiol.* Vol. 5, No. 4, pp. 429–440, March 20, 1923.
19. Relation between time-drop and serum antibodies. *J. Exp. Med.* Vol. 37, pp. 659–669, May 1, 1923.
20. The study of immune serum. Time-drop and initial value of surface tension. *J. Exp. Med.* Vol. 38, pp. 87–92, July 1, 1923.
21. Signification de la chute maxima de tension superficielle du serum sanguin. *C. R. Ac. Sc.* t. 177, p. 1140, November 1923.
22. Chute spontanée de la tension superficielle du sérum et de ses solutions. *C. R. Soc. Biol.* t. 89, p. 1015, November 1923.
23. Tension superficielle du sérum. Relation entre la chute en fonction du temps et des anticorps. *C. R. Soc. Biol.* t. 89, p. 1146, December 1923.
24. Les phénomènes de tension superficielle en biologie. *C. R. Soc. Biol.* t. 89, p. 1076, December 1923.
25. Tension superficielle du sérum. Chute de la tension superficielle due à l'addition de certaines substances et action antagoniste du sérum. *C. R. Soc. Biol.* t. 89, p. 1148, December 1923.
26. Significance of the maximum time-drop of serum solutions. *J. Exp. Med.* Vol. 39, pp. 37–41, January 1, 1924.
27. Sur les variations de la viscosité du sérum sanguin en fonction de la température. *C. R. Soc. Biol.* t. 90, No. 3, pp. 168–170, February 1, 1924.
28. Dimensions des molécules de certaines substances colloidales. *C. R. Ac. Sc* t. 178, pp. 1102, March 1924.
29. Au sujet d'une couche monomoléculaire absorbée sur les globules rouges et les parois des capillaires. *C. R. Soc. Biol.* t. 15, p. 1450, May 1924.
30. Further evidence indicating the existence of a superficial polarized layer of molecules at certain dilutions. *J. Exp. Med.* Vol. 39, pp. 717–724, May 1924.
31. The surface equilibrium of colloidal solutions and the dimensions of some colloidal molecules. *Science.* Vol. 59, No. 1539, pp. 580–582, June 1924.
32. Dimensions des molécules et poids moléculaires des protéines du sérum. *C. R. Ac. Sc.* t. 178, p. 1904, June 1924.

33. Time-drop and smallpox vaccination. *J. Exp. Med.* Vol. 40, pp. 129–132, July 1, 1924.
34. On the thickness of the monomolecular layer of serum. *J. Exp. Med.* Vol. 40, pp. 133–149, July 1, 1924.
35. An improvement of the technique for measuring surface tension. *J. Gen. Physiol.* Vol. 6, pp. 625–628, July 20, 1924.
36. Surface tension of colloidal solutions and dimensions of certain organic molecules. *Philos. Mag.* Vol. 48, pp. 264–277, August 1924.
37. The surface equilibrium of colloidal solutions. Antagonistic action of colloids. *Science.* Vol. 60, No. 1554, pp. 337–338, October 1924.
38. A new determination of the constant *N* of Avogadro, based on its definition. *Philos. Mag.* Vol. 48, No. 286, pp. 664–672, October 1924.
39. Die Bestimmung der Oberflächenspannung mit der Ring-Methode (Torsionswage). *Biochem. Zeit.* Bd. 155, Heft½, December 1924.
40. A note on the surface viscosity of colloidal solutions *Science.* Vol. 61, No. 1570, p. 117, January 30, 1925.
41. Une méthode physique aussi sensible que les réactions anaphylactiques pour déceler des traces de protéines en solutions. *C. R. Soc. Biol.* t. 92, p. 1194, May 1925.
42. A technique for accurate study of the drop in function of the time. *J. Exp. Med.* Vol. 41, pp. 663–672, May 1, 1925.
43. Une novelle méthode de détermination des dimensions moléculaires et du nombre *N* basée sur l'étude de l'équilibre superficiel de solutions colloidales. *J. Phys. et Radium.* ser. 6, t. 6, No. 5, pp. 145–153, May 1925.
44. An interfacial tensiometer for universal use. *J. Gen. Physiol.* Vol. 7, No. 5, pp. 625–631, May 20, 1925.
45. Appareil pour la measure rapide de la tension superficielle à la surface de séparation de deux liquides. Influence de la température. *C. R. Ac. Sc.* t. 180, p. 1579, May 25, 1925.
46. A highly sensitive physical method for detecting proteins in a solution. *Science.* Vol. 41, No. 1583, p. 472, May 1925.
47. Modification physico-chimique momentanée du sérum consécutive à l'injection d'antigène. *C. R. Soc. Biol.* t. 93, p. 14, June 1925.
48. On certain physico-chemical changes in serum as a result of immunization. *J. Exp. Med.* Vol. 41, pp. 779–793, June 1, 1925.
49. Some new aspects of the surface tension of colloidal solutions which have led to the determination of molecular dimensions. 3rd Colloid Symposium (St. Paul, Minn.), June 1925.
50. On the probable dimensions of the molecule and molecular

weight of crystalline egg albumin. *J. Biol. Chem.* Vol. 44, No. 3, pp. 595–613, July 1925.

51. Concerning the change in surface tension occurring as a result of immunization. *J. Exp. Med.* Vol. 42, pp. 9–15, July 1, 1925.
52. The constant *N* of Avorgrado, correction. *Philos. Mag.* Vol. 1, p. 664, August. 1925.
53. Die Oberflächenspannung von Serum (Über physikalisch-chemische Änderungen im Serum als Folge der Immunisierung). *Biochem. Zeit.* Bd. 165, Heft ⅓, p. 134, September 1925.
54. Studi sperimentali sulla tensione superficiale del siero. Biochemicale Terapia Sperimentale. An. Dodicessimo, Fasc. Sec., 1925.
55. Recherches sur l'équilibre superficiel des solutions colloidales. Application de la méthode à la détermination des dimensions de certaines molécules. C. R. Congres, Grenoble, 1925.
56. De Démocrite à Einstein. La psychologie de la recherche scientifique. Lecture, Montreal, 1925.
57. Au sujet d'une note de Goiffon sur la mesure de la tension superficielle. *C. R. Soc. Biol.* t. 94, p. 266, January 30, 1926.
58. An hypothesis on cell structure and cell movements based on thermodynamical considerations. *Science.* Vol. 43, No. 1628, pp. 284–286, March 12, 1926.
59. Surface tension of colloidal solutions and dimensions of certain organic molecules. In Alexander, *Colloid Chemistry.* New York, 1926, p. 267.
60. Cytological measurements to test du Noüy's thermodynamical hypothesis of cell size. *Anat. Record.* Vol. 34, No. 5, p. 313, February 1927.
61. Concerning the ring method for measuring surface tension. *Science.* Vol. 65, No. 1676, pp. 160–161, February 1927.
62. Sur la viscosité du sérum sanguin en fonction de la température. *C. R. Soc. Biol.* t. 96, p. 1203, May 1927.
63. Sur la nature de dispersion des substances constituant le plasma et le sérum et sur les dimensions possibles de la "Molecule de Plasma." *C. R. Soc. Biol.* t. 96, p. 1205, May 1927.
64. Sur une anomalie dans la vitesse d'évaporation des solutions d'oléate de soude et de digitonine aux dilutions élevées. *C. R. Ac. Sc.* t. 184, p. 1062, May 1927.
65. The thickness of the monolayer of rabbit plasma. *J. Exp. Med.* Vol. 46, p. 16, July 1, 1927.
66. A densimeter for the rapid determination of the specific gravity of small quantities of liquids and solids. *J. Biol. Chem.* Vol. 74, No. 3, p. 443, September 1927.
67. Sur une modification spontanée de la viscosité du sérum sanguin. *C. R. Ac. Sc.* t. 186, p. 804, March 1928.

68. A spontaneous modification of the viscosity of blood serum. *Science.* Vol. 67, No. 1744, pp. 563–564, June 1928.
69. Etude sur la viscosité du sérum sanguin en fonction de la température et sur l'hydradation de ses protéines. *Ann. Inst. Pasteur.* t. 42 (ler mémoire), p. 742, July 1928.
70. Sur la capacité d'absorption des protéines des sérum vis-à-vis des sels biliaires. *C. R. Soc. Biol.* t. 99, p. 1097, October 13, 1928.
71. Some physico-chemical characteristics of immune serum. In Alexander, *Colloid Chemistry* II, New York, 1928, p. 775.
72. The viscosity of blood serum as a function of temperature. *J. Gen. Physiol.* Vol. 12, No. 3, pp. 363–377, January 1929.
73. L'evolution des sciences de la vie. *Revue de Paris.* January 1929.
74. Sur l'indice de réfraction, le coefficient de température et la dispersion du sérum sanguin. *C. R. Soc. Biol.* t. 100, p. 490, February 16, 1929.
75. Sur le pouvoir rotatoire du sérum en fonction de la température. *C. R. Ac. Sc.* t. 188, p. 660, February 1929.
76. A device for measuring surface tension automatically. *Science.* Vol. 69, No. 1783, pp. 251–252, March 1929.
77. On the rotatory power of serum. *Science.* Vol. 69, No. 1795, pp. 552–553, May 1929.
78. Etude sur le pouvoir rotatoire et la dispersion rotatoire du sérum en fonction du temps et de la température. *Ann. Inst. Pasteur.* t. 43 (2éme mémoire), p. 749, June 1929.
79. Recherches sur la température critique du sérum (55°–56°) au moyen des mesures photométriques. *C. R. Soc. Biol.* t. 101, p. 359, June 1929.
80. Etudes sur les phénomèmes de tension superficielle des solutions colloidales biologiques. *Protoplasma.* t. 6, No. 3, p. 494, 1929.
81. Recherches sur la température critique du sérum (55°–56°) au moyen des mesures photométriques. *Ann. Inst. Past.* t. 44 (3ème mémoire), p. 109, January 1930.
82. Untersuchungen über die Oberflächenspannung des Serums. Lecture, Vienna, January 1930.
83. Transmission and diffraction of light by normal serum as a function of temperature. *Science.* Vol. 71, No. 30, p. 108, January 1930.
84. La biologie moderne et le problème de la vie. *Revue Scientifique.* February 8, 1930.
85. Température critique du sérum. Mesure du facteur de dépolarisation en fonction de la température. *C. R. Soc. Biol.* t. 104, p. 146, May 1930.
86. Sur la température critique du sérum et la coagulation du

sérum par la chaleur. *C. R. Soc. Biol.* t. 106, p. 289, May 17, 1930.

87. Sur le mécanisme de la coagulation du sérum par la chaleur. *C. R. Soc. Biol.* t. 106, p. 377, May 24, 1930.
88. Recherches sur la température critique du sérum. Mécanisme de la coagulation, etc. *Ann. Inst. Pasteur.* t. 45 (4ème mémoire), p. 251, August 1930.
89. On the critical temperature of blood serum. Depolarization factor and hydration of serum molecules. *Science.* Vol. 72, No. 1861, pp. 224–225, August 29, 1930.
90. Recherches sur les équilibres ioniques du sérum. *C. R. Soc. Biol.* t. 106, p. 85, January 17, 1931.
91. La tension superficielle des liquides de l'organisme. Conférence à Nancy, Fac. de Médecine, January 1931.
92. L'action des rayons *X* sur les cultures de tissus in vitro (Avec Trillat et autres.) *C. R. Ac. Sc.* t. 192, p. 304, February 1931.
93. Recherches sur les équilibres ioniques du sérum (Avec V. Hamon). *C. R. Soc. Biol.* t. 106 (2ème note), p. 352, February 1931.
94. La mesure de la viscosité. Un viscosimètre de précision. *Revue Scientifique.* February 28, 1931.
95. Replacing the telephone by a loud speaker for measurements of conductivity. *Nature.* Vol. 127, No. 3203, p. 441, March 1931.
96. On a tilting gas stopcock. *Science.* Vol. 73, No. 1898, p. 530, May 1931.
97. La tension superficielle et sa mesure. *La Nature.* No. 2857, p. 435, May 15, 1931.
98. Ionic equilibria in the serum in relation to the critical temperature. *Science.* Vol. 73, p. 595, May 1931.
99. Les cultures de tissus en dehors de l'organisme. *Revue de Paris.* p. 117, September 1, 1931.
100. La bactériologie nouvelle. *Le Correspondant.* September 25, 1931.
101. Le problème biologique et la valeur des méthodes. Bruxelles Médical (conference), No. 48, p. 65, September 27, 1931.
102. Effect of light on the surface tension of Boy's soap solution. *Nature.* Vol. 128, No. 3233, p. 496, October 1931.
103. Spectrophotométrie du sérum dans le visible et le proche infrarouge (Avec M. Lecomte du Noüy). *C. R. Soc. Biol.* t. 108, p. 657, November 1931.
104. Mesure de la concentration en ions hydrogène des liquides au moyen d'une électrode rotative. *C. R. Ac. Sc.* t. 193, p. 1417, December 1931.
105. Les cultures de tissus en dehors de l'organisme. *Bull. de la Société Philomathique.* t. 20, série 10, p. 39, 1931.

106. La Bactériologie nouvelle. *Arch. Institut Prophylactique.* t. 4, No. 1, 1932.
107. Surface tension of soap solutions. *Nature.* February 1932.
108. Recherches sur la température critique du sérum. Equilibres ioniques du sérum en fonction de la température. *Ann. Inst. Pasteur.* t. 48 (5ème mémoire), p. 187, February 1932.
109. Electrode rotative pour la mesure du pH sanguin (Avec Loiseleur). *C. R. Soc. Biol.* t. 109, p. 1181, April 1932.
110. Sur la tension superficielle des liquides de l'organisme. *C. R. Soc. Biol.* t. 109, p. 1069, April 1932.
111. Une mesure de l'activité physiologique. *C. R. Soc. Biol.* t. 109, p. 1227, April 1932.
112. Etudes sur la température critique du sérum. Spectre d'absorption du sérum de cheval dans l'ultra-violet (Avec M. Lecomte du Noüy). *C. R. Ac. Sc.* t. 194, p. 1815, May 23, 1932.
113. Sur un sérum pathologique incoagulable par la chaleur (P. et M. Lecomte du Noüy et M. Aynaud). *C. R. Soc. Biol.* t. 110, p. 333, June 1932.
114. A new hydrogen electrode and apparatus for the determination of pH. *Science.* Vol. 75, No. 1955, pp. 643–644, June 1932.
115. Sur la température critique du sérum (55°–57°). Phénomènes ioniques. La conductivité du sérum normal et immunisé en fonction de la température. *C. R. Soc. Biol.* t. 110, p. 485, June 1932.
116. The biological problem. *Emmanuel Libman Anniversary Volumes.* New York, 1932.
117. L'étude des propriétés physiques et physico-chimiques du sérum. *Soc. de Pathologie comparée.* Nov. 10, 1932.
118. Recherches sur la température critique du sérum. Le spectre d'absorption dans l'ultra-violet, le visible et le proche infrarouge (Avec M. Lecomte du Noüy). *Ann. Inst. Pasteur.* t. 99 (6ème mémoire), p. 762, December 1932.
119. Perfectionnements à l'électrode d'hydrogène pour la mesure de la concentration en ions hydrogène des solutions. *C. R. Ac. Sc.* t. 195, p. 1265, December 1932.
120. Recherches sur le sérum et sa température critique. La conductivité électrique du sérum en fonction de la température. *Ann. Inst. Pasteur.* t. 50 (7ème mémoire), p. 127, January 1933.
121. La sériologie à la lumière des données nouvelles sur les propriétés physico-chimiques des sérums. *Soc. de Sériologie.* January 12, 1933.
122. Spectrophotométrie du sérum dans l'ultra-violet. Sur une altération profonde des protéines suivie de mort (Avec M. Lecomte du Noüy). *C. R. Soc. Biol.* t. 112, p. 1267, April 1933.

123. Surface tension of colloidal solutions, and the action of light on soap solutions. *Nature*. Vol. 131, No. 3315, p. 689, May 1933.
124. Le sérum sanguin. *Archives Inst. Prophylactique*, 2ème trimestre, 1933.
125. Phénomènes physico-chimiques dans l'immunité (rapport). Congrès de Rome, September 1933.
126. Le Vieillissement. *Revue de Paris*. November 1933.
127. Sur la mesure du pH du plasma sanguin. Etude expérimentale sur l'électrode rotative inclinée (Avec V. Hamon). *Bull. Soc. Chimie Biol.* t. 16, No. 2, p. 177, 1934.
128. Les aspects physico-chimiques de l'immunité. *Ergebn. der Hygiene bakteriol.* Immunit. m. Ther, Bd. 15, p. 304, 1934.
129. Appareillage technique pour la mesure du pH. 3ème Congrès international des Indust. Agricoles, Paris, 1934.
130. Contribution à l'étude des bases chimiques de la classification naturelle (Avec M. Lecomte du Noüy). *C. R. Soc. Biol.* t. 116, p. 108, 1934.
131. The pH of serum inactivated by heat. *Nature*. Vol. 134, p. 628, October 1934.
132. Le pH du sérum inactivé par la chaleur (Avec V. Hamon). *C. R. Soc. Biol.* t. 117, p. 337, October 20, 1934.
133. Les aspects physiques des réaction de floculation. Conférence à la Soc. Franc. de Sériologie et de Syphilis expérimentale, November 8, 1934.
134. Ring method for measuring surface tension. *Nature*. Vol. 135, p. 397, March 1935.
135. Sur une nouvelle methode de dosage de l'antitoxine diphtérique, par la viscosité (Avec V. Hamon). *C. R. Ac. Sc.* t. 200, p. 1250, 1935.
136. Recherches sur la température critique du sérum. Equilibres ionique en fonction de la température: la pH (Avec V. Hamon). *Ann. Inst. Pasteur.* t. 54, p. 442 (9ème mémoire), 1935.
137. La lutte pour la vie. Conférence radiodiffusée. January 23, 1935.
138. Le laboratoire et la recherche scientifique. *Revue de Paris*. June 1935.
139. Immunological reactions and viscosity. *Science*. Vol. 82, No. 2124, p. 254, September 13, 1935.
140. Biochimie et biophysique (rapport). Vème Congrès de Chimie Biologique, Bruxelles, October 1935.
141. Les mirages de la science. *Revue de Paris*. March 1, 1936.
142. Biophysique moléculaire et immunologie. Livre d'or du Prof. Roffo, Buenos Aires, 1936.
143. La viscosité des mélanges, toxine-antitoxine diphtérique (Avec V. Hamon). *Ann. Inst. Pasteur.* t. 56, p. 359, April 1936.

144. Fortschritte der biologischen Wissenschaften in Frankreich. *Geistige Arbeit*, Berlin. No. 13, July 1936.
145. Le temps et la vie. Conférence radiodiffusée. Ext. *Revue "Science."* 1936.
146. Considérations expérimentales sur l'absorption en biologie. Conférence à la Soc. de Chimie Biologique. t. 18, No. 5, May 1936.
147. Sur l'unité de la méthode dans les sciences physiques et biologiques. Actes du Congrès International de Philosophie Scientifique, Sorbonne, Paris, 1935–1936.
148. Réflexions sur l'organisation du laboratoire de recherche. *Chimie et Industrie*. Vol. 36, No. 2, August 1936.
149. Le laboratoire moderne de recherche. *Chimie et Industrie*. Vol. 87, No. 1, January 1937.
150. Das Kausalproblem-Diskussionsbemerkungen zum Vortag Franks. II Internat. Kongress fur Einheit der Wissenschaft, Erkenntnis, Bd. 6, Heft 5/6, Copenhagen, 1936.
151. Das Kausalproblem-Diskussionsbemerkungen zum Vortag Rashevekys. Copenhagen, 1936.
152. The critical temperature of blood serum (The behavior of serum proteins as a function of the temperature and the mechanism of coagulation). *C. R. Lab.* Carlsberg, Ser. Chi. Vol. 22, jubilaire du Prof. Sørensen, October 5, 1937.
153. Filtration. Extrait de "Les Ultravirus des Maladies Humaines," sous la direction de MM. C. Levaditi et P. Lépine, Lib. Maloine, Paris, 1938.
154. L'absorption dans les phénomènes biologiques. *Livro de Homenagen*. pp. 398–404, Rio de Janeiro, 1939.
155. Das Altern und die physiologische Zeit. Zeitsch, f. Altersforschung, Bd. I, Heft 4, January 1939.
156. Notre univers et son image. *Revue Genérale des Sciences*. G. Doin & Cie, Paris, May 1939.
157. Au sujet d'un art, de Mr. Kopaczewski sur "La composition et la structure du sérum" publié dans la *Revue Générale de Sciences*, du 15 Mai 1939. *Revue Générale des Sciences*. G. Doin & Cie, Paris, June 15, 1939.
158. Physique moléculaire. Une nouvelle méthode d'études des huiles de graissage et de leurs propriétés. *C. R. Ac. Sc.* t. 210, pp. 101–102, January 15, 1940.
159. Sur quelques faits nouveaux concernant l'équilibre superficiel de solutions complexes. *C. R. Ac. Sc.* t. 210, pp. 334–335, February 26, 1940.
160. A new method for studying the properties of lubricating oils based on the use of a new instrument. *Science*. Vol. 91, No. 2362, pp. 341–342, April 5, 1940.

161. Un nouvel instrument de recherche et de contrôle: le tensiomètre enregistreur. *Chimie et Industrie.* Vol. 43, No. 3, February 5, 1940.
162. Au sujet d'un mémoire de MM. Rubinstein et Kusimina sur le "Phénomène de du Noüy." *Chim. Biol.* t. 22, Nos. 7–8, July-August 1940.
163. A propos de la communication de MM. A. Dognon et L. Gougerot intitulée "Enregistrement des tensions de surface. Evolution des surfaces et signification du phénomène de du Noüy." *Bull. Soc. Chim. Biol.* t. 23, Nos. 4–6, April and June 1941.
164. L'institut de recherche de demain. *Chimie et Industrie.* Vol. 46, No. 5, pp. 709–713, November 1941.
165. Physiological time. *Am. Philos. Soc.* Vol. 87, No. 5, 1944.
166. La France inspiratrice. Quelques grands inventeurs. *Les Cahiers:* Le Bayou, 1946.

INDEX